Introduction to Chaos

Introduction to Chaos

Physics and Mathematics of Chaotic Phenomena

Hiroyuki Nagashima and Yoshikazu Baba

Shizuoka University,
Shizuoka, Japan

Translated from Japanese by

Mikio Nakahara

Kinki University,
Higashi-Osaka, Japan

Institute of Physics Publishing
Bristol and Philadelphia

British Library Cataloguing-in-Publication Data

A catalogue record for this book is available from the British Library.

ISBN 0 7503 0507 X (hbk)
0 7503 0508 8 (pbk)

Library of Congress Cataloging-in-Publication Data are available

This English edition has been translated and revised from the original Japanese publication, *Introduction to Chaos*, edited by Hiroyuki Nagashima and Yoshikazu Baba. © 1992 Baifukan Co., Ltd, Japan.

Published by Institute of Physics Publishing, wholly owned by The Institute of Physics, London

Institute of Physics Publishing, Dirac House, Temple Back, Bristol BS1 6BE, UK

US Office: Institute of Physics Publishing, The Public Ledger Building, Suite 1035, 150 South Independence Mall West, Philadelphia, PA 19106, USA

Typeset in TEX using the IOP Bookmaker Macros
Printed in the UK by J W Arrowsmith Ltd, Bristol

Contents

Preface to the first (Japanese) edition

This book is an introduction to chaos. It is written primarily for advanced undergraduate students in science but postgraduate students and researchers in mathematics, physics and other areas of science will also find this book interesting.

Instead of exhausting all the topics in chaos, we explained this theory by taking examples from one-dimensional maps and simple differential equations. In due course, we also explained some elementary mathematical physics. Since this book is meant to be an introduction, we gave numerous diagrams and computer graphics in the text to facilitate readers' understanding.

This book is a result of a collaboration between a mathematician (YB) and a physicist (HN). Chapter 2 was written mainly by YB while HN wrote the rest of the book. We did not, however, write these parts separately but we kept examining and criticizing each other's manuscripts. In this sense, it is quite right to say that this is the result of our collaboration. While we were writing this book, we had to adjust the differences between our ways of thinking or our ways of expression many times, due to our differing viewpoints as a physicist and a mathematician. We also found numerous new problems while writing this book and, in fact, we made a new discovery on a one-dimensional map. As a result of these delights, our writing proceeded slowly. We would like to thank Mr Takashi Murayama of Baifukan Publishing Co. for his patience.

The diagrams in the text were mostly drawn by HN using a HP 9000 workstation. Some diagrams were prepared by M Ohba, I Itoh and H Kayama who carried out their undergraduate projects at HN's lab. The double pendulum referred to in the text was made by our technician, Mr K Masuda. Professor T Asai of the Mathematics Department, Shizuoka University, made many valuable suggestions concerning number theory, which helped us a great deal. Mr Seiji Iwata of Baifukan did an excellent job in editing our manuscript. Finally we would like to express our sincere thanks to all the members of the Science Section, Faculty of Liberal Arts, Shizuoka University.

Hiroyuki Nagashima (Deceased)
Yoshikazu Baba
July 1992

Preface to the English edition

Professor Hiroyuki Nagashima, one of the leading physicists in chaos, both theoretical and experimental, passed away in Shizuoka on 4 February 1998. His early death is an enormous loss to research in chaos. All of his friends miss his vigour and enthusiasm.

Although Professor Nagashima wanted to add appendices on singularity, his illness made it impossible. Professor Baba completed these appendices based on the notes and references that Professor Nagashima had left. Thus appendices 4E and 4D on singularity have been added in this English edition. A large number of references have also been added. They are indicated as [4], for example, and can be found in the reference list.

A few additional notes on this translation are in order. There are several BASIC programs in the original Japanese edition. They were programmed for a PC that understands the Japanese language system. Therefore they have been replaced by more widely available *Mathematica* programs by me. Accordingly I have produced figures 4C.2 and 4D.1.

I would like to thank Professors H Nagashima and Y Baba for their assistance while preparing this translation. Several TeX-nical problems have been solved by Al Troyano, Yukitoshi Fujimura and Shin'ichiro Ogawa. Their assistance is greatly acknowledged. Finally, I would like to thank Jim Revill of Institute of Physics Publishing for his patience.

Mikio Nakahara
October 1998

Chapter 1

What is chaos?

The meaning or the definition of chaos must be clarified, first of all, to discuss chaos. Instead of stating these in lexicographic order, they are explained in due course as concrete examples are introduced.

1.1 What is chaos?

The word 'chaos' originates from a Greek word Χάος and its everyday meaning is 'a state without order'. This word reminds one of a totally disorganized state, in contrast with *cosmos*, an ordered state. Although 'chaos' in this book is not an ordered state, it indicates a phenomenon which is not totally disordered but disordered moderately and shows a temporarily irregular motion. Moreover, it refers to

> 'an irregular oscillation governed by a relatively simple rule'.

Here 'oscillation' simply represents a variation of certain quantities.

The reader must wonder if there exists an irregular oscillation governed by a simple rule at all. Figure 1.1 serves as evidence for this statement. This is a graph of an irregular series $\{x_n\}$ as a function of n. Where on earth is a simple rule hiding behind this irregular series? Let us 'discover' the regularity of this series as follows.

Suppose, first of all, that the $(n+1)$th term x_{n+1} depends only on the nth term x_n. This may be written as the *Ansatz*

$$x_{n+1} = f(x_n). \tag{1.1}$$

To visualize this relation concretely, let us consider a plane whose Cartesian coordinates are x_n (abscissa) and x_{n+1} (ordinate). That is, the points (x_n, x_{n+1}) $(n = 1, 2, 3, \ldots)$ are plotted in order. This plot reveals a parabola, which is convex upward, as shown in figure 1.2. This relation is called a *map* from x_n to x_{n+1}. The map in our case is

$$x_{n+1} = L(x_n) = 4x_n(1 - x_n). \tag{1.2}$$

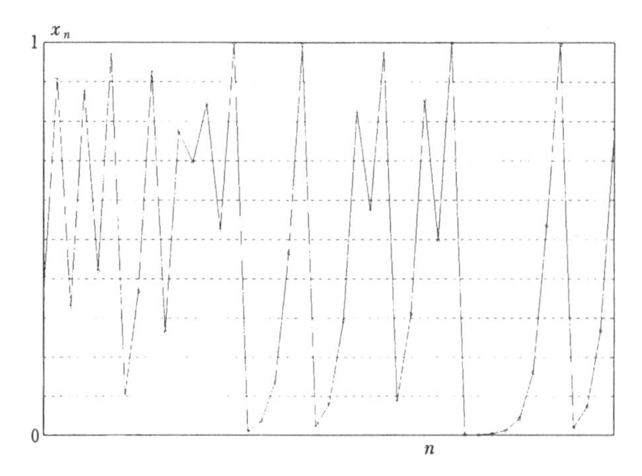

Figure 1.1. An irregular series.

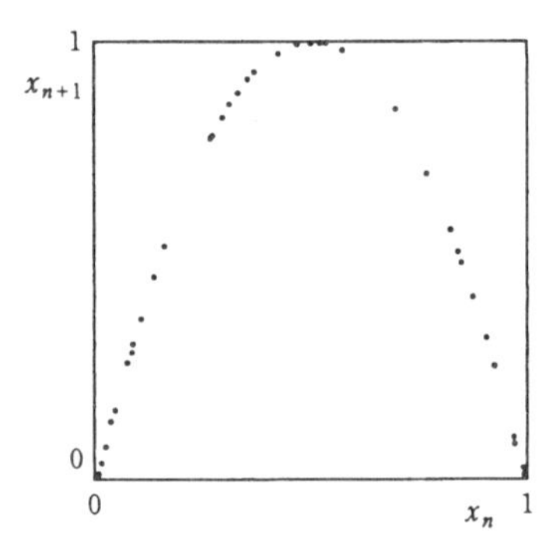

Figure 1.2. The plot of (x_n, x_{n+1}) $(n = 1, 2, 3, \ldots)$ for the series shown in figure 1.1. This plot defines the logistic map.

In other words, the function $f(x)$ in equation (1.1) is given by $f(x) = 4x(1-x)$. The relation where the value of x_{n+1} is determined only by x_n is called traditionally a *one-dimensional map*. The map defined by equation (1.2) is called the *logistic map*. It is surprising that a series generated by such a simple rule looks apparently irregular. The irregularity of this series is better seen if it is plotted to larger n than those plotted in figure 1.1. The series then behaves as if it takes almost any value in the interval [0, 1]. It is a phonomenological peculiarity of chaos that it behaves irregularly even though it is generated by such a simple rule as equation (1.2).

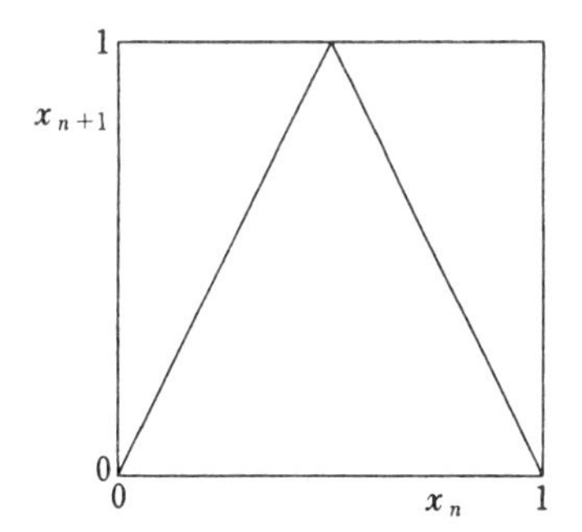

Figure 1.3. The tent map.

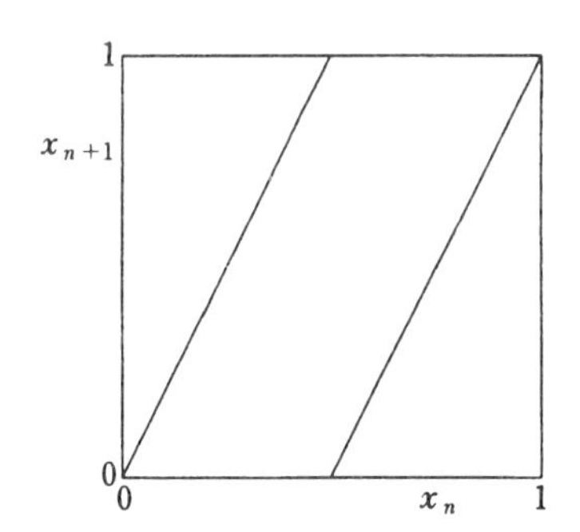

Figure 1.4. The binary transformation, also known as the Bernoulli shift.

There are infinitely many simple rules, besides equation (1.2), which generate irregular series. Two typical examples are

$$x_{n+1} = T(x_n) = \begin{cases} 2x_n & (0 \le x_n \le 0.5) \\ 2 - 2x_n & (0.5 < x_n \le 1) \end{cases} \tag{1.3}$$
$$= 1 - |1 - 2x_n|$$

$$x_{n+1} = B(x_n) = \begin{cases} 2x_n & (0 \le x_n \le 0.5) \\ 2x_n - 1 & (0.5 < x_n \le 1). \end{cases} \tag{1.4}$$

The variable x_n in equations (1.2), (1.3) and (1.4) should be restricted in the interval [0, 1].

The map (1.3) is called the *tent map*, while (1.4) the *binary transformation* or the *Bernoulli shift*. The graphs of these maps are shown in figures 1.3 and 1.4 respectively.

Problem 1. Determine which of the one-dimensional maps (1.2)–(1.4) generates the irregular series shown in figures 1.5–1.7. Plot (x_n, x_{n+1}) for identification.

It might seem rather mysterious that these simple maps generate irregular sequences. To understand this mystery one should run the following Mathematica program to generate an irregular series,

```
x=0.3
For[k=1, k<=50, k++, x=4 x (1-x), Print[x]]
```

for example.[1]

One then inputs a number in the interval [0, 1] as an initial value of x (=0.3 in the example above). Then the series $\{x_n\}$ is irregular, unless the initial value is a very special one.[2]

[1] This is programmed in BASIC language in the original text. This and subsequent Mathematica programs are provided by the translator.

[2] The special value yields a periodic series and there are an infinite number of them in the interval [0, 1]. Fortunately most of them are irrational for the logistic map (1.2), see appendix 1A.

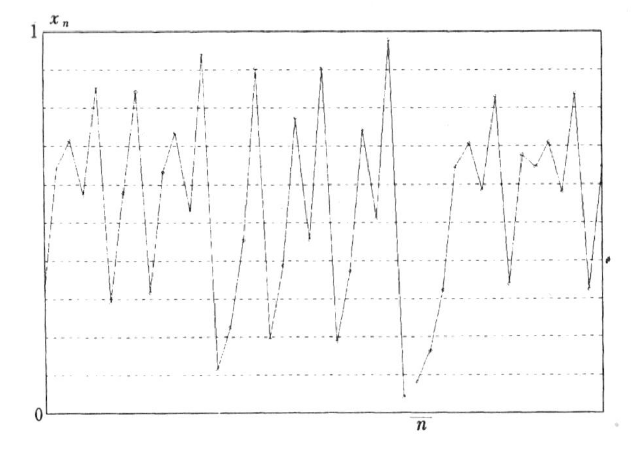

Figure 1.5.

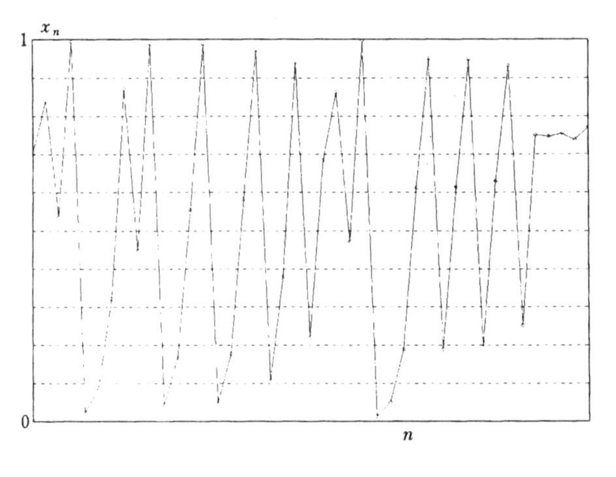

Figure 1.6.

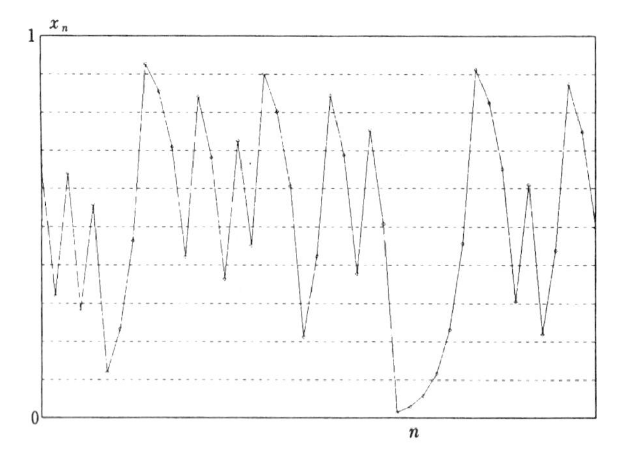

Figure 1.7.

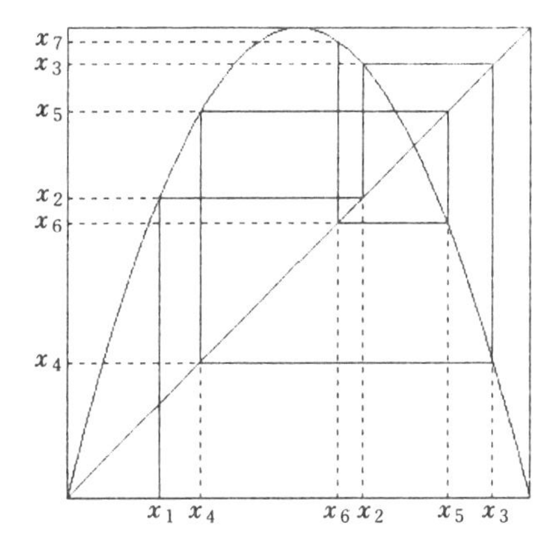

Figure 1.8. How to obtain the series $\{x_n\}$ by making use of the graph $x_{n+1} = L(x_n) = 4x_n(1 - x_n)$. An initial value x_1 generates $x_2, x_3, \ldots$ iteratively. It turns out, as shown in the figure, that this series is easily obtained by an alternative use of $x_{n+1} = L(x_n)$ and $x_{n+1} = x_n$.

A more direct method than the Mathematica program above is to use a graph. The reader should be referred to figure 1.8 to find how to obtain the series from the graph.

1.2 Characteristics of chaos

What are the characteristics of the irregular behaviour generated by the maps (1.2)–(1.4)? One may readily find from figures 1.2–1.4 that the graph of x_{n+1} as a function of x_n is twofold, that is, there are two x_n values which yield a given x_{n+1}. Then the inclination of the graph is necessarily steep on average. Steep inclination implies that two series that started off with very close initial conditions are separated quickly as iterated. This can be observed in figure 1.9.

Let us see this explicitly. Figure 1.10 shows two series generated by the Mathematica program (1.5) with the initial conditions $x_0 = 0.35$ and $x_0' = 0.350\,001$. It shows that the two series are separated abruptly as n becomes larger. Thus the results of the iterations are very different even if the initial conditions are only slightly different. This *sensitivity to the initial condition* is a characteristic of chaos.

To see how sensitive the result is, it is more convenient to analyse equations (1.3) and (1.4), for which the absolute value of the inclination is a constant 2, rather than the logistic map (1.2). Let us take initial values x_0 and $x_0 + \Delta x$. The difference Δx becomes $f'(x_0)\Delta x = 2\Delta x$ after an iteration. Suppose Δx is as small as 10^{-5} for example. Then, after 15 iterations, the difference is $\Delta x 2^{15} = 10^{-5}2^{15} = 0.33$. As the distance of the points approaches 0.5 in equations (1.3) and (1.4), it becomes very likely that these points are separated

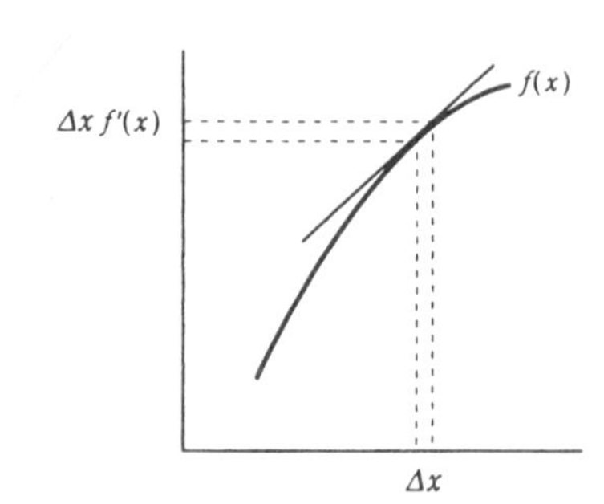

Figure 1.9. The expansion of an interval Δx after an iteration.

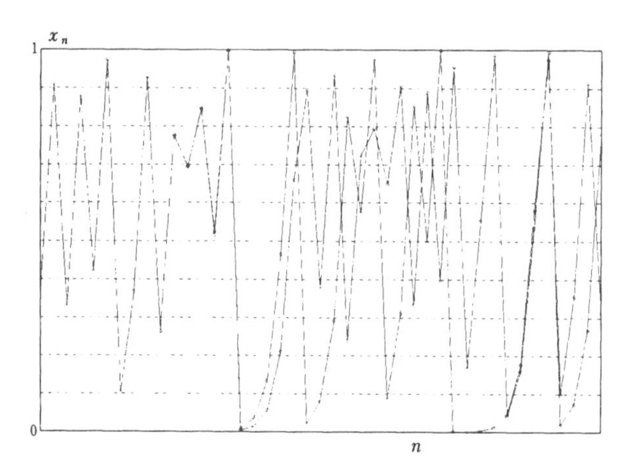

Figure 1.10. Two series with very close initial conditions $x_0 = 0.35$ and $x_0' = 0.350\,001$.

on different sides of the point $x_n = 0.5$. When this is the case, the mutual distance is not multiplied by 2 any longer on iteration, and these points behave quite differently from then on in spite of the very close initial conditions. Let us consider the fate of the interval $[x_0, x_0']$ whose end points are two initial values. This interval is elongated twice as much on each iteration in the beginning but eventually it is folded. The operation of 'elongation' followed by 'folding' is the most fundamental property of a map generating chaos.

Let us consider the baker's transformation to familiarize ourselves with the operations of elongation and folding. The term 'baker's transformation' is coined after the analogy of the operation with the way dough is prepared when a baker makes a pie. This transformation is a map of a two-dimensional figure defined in figure 1.11.

An interval is stretched double horizontally, then cut into half and piled

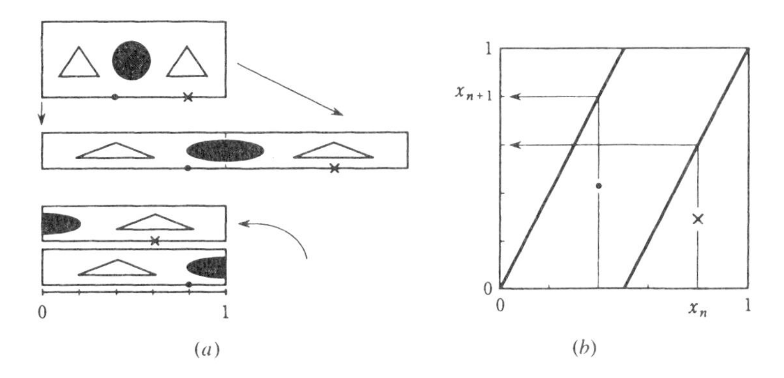

Figure 1.11. The relation between (a) the baker's transformation and (b) the binary transformation. The initial disc is elongated horizontally if the baker's transformation is repeatedly applied.

up as shown in figure 1.11.[3] If this operation is repeated many times, the disc in figure 1.11 spreads all over the domain and the horizontal distribution of the disc becomes uniform. Therefore the horizontal change is just the transformation defined by equation (1.4).

Problem 2. Show graphically the two-dimensional transformation whose horizontal change is the transformation (1.3).

It turns out that the ingredients of pie separated in the beginning spread throughout the dough and are mixed up by the repeated applications of the baker's transformation. (Note that this is true only horizontally. The mixing does not exist in the vertical direction, along which the layered structure is piled up.) If one chooses a point in [0, 1], which is not one of the periodic points, as the initial value of the series $\{x_i\}$ generated by the maps (1.2)–(1.4), then x_i are distributed almost all over this interval. This fact will be studied in detail in chapter 3. Here we only show the distribution of the numbers in the irregular series generated by the logistic map and the tent map in figures 1.12 and 1.13, respectively. It can be seen from these figures that the series distribute throughout the interval [0, 1].

It should be noted that the sensitivity to the initial value suggests the irregularity of the series $\{x_i\}$ generated by the maps previously defined. Let

[3] This operation φ is expressed mathematically as

$$\varphi(x, y) = \begin{cases} \left(2x, \frac{1}{2}y\right) & \left(0 \le x \le \frac{1}{2}\right) \\ \left(2x - 1, \frac{1}{2}(y+1)\right) & \left(\frac{1}{2} < x \le 1\right). \end{cases}$$

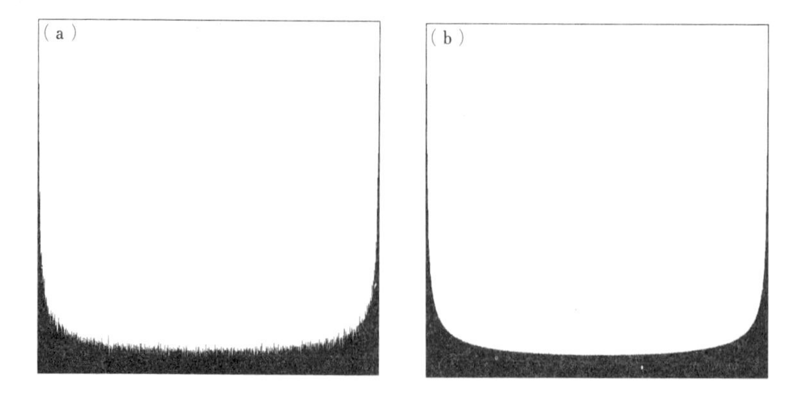

Figure 1.12. The distribution of the numbers in the series generated by the logistic map. The map is iterated (a) 10^5 times and (b) 10^7 times. The interval [0, 1] is divided into 1000 small intervals.

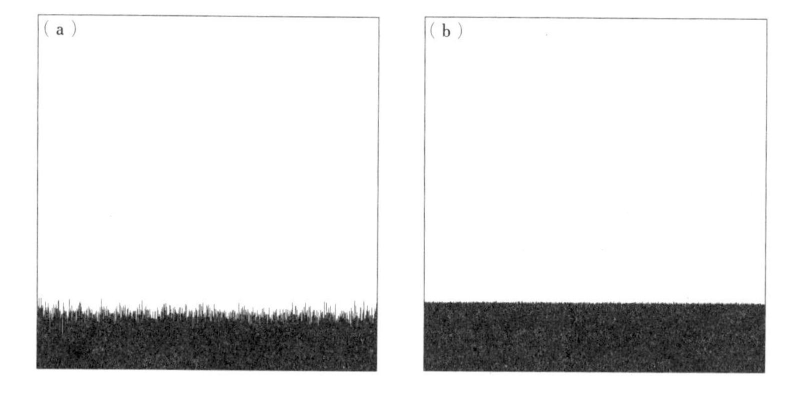

Figure 1.13. The distribution of the numbers in the series generated by the tent map. The interval [0, 1] is divided into 1000. (a) The map is iterated 10^5 times and the fluctuation is approximately $1/\sqrt{10^5/10^3} = 0.1$. (b) The map is iterated 10^7 times and the fluctuation is approximately 0.01.

us concentrate on the ith number x_i of the series to see this. It may be possible that x_j with $j > i$ is quite close to x_i. Unless $x_i = x_j$ exactly, however, the part $x_i, x_{i+1}, x_{i+2}, \ldots$ is very different from the part $x_j, x_{j+1}, x_{j+2}, \ldots$ due to the sensitivity to the initial condition.

Problem 3. Suppose the program (1.5) is executed on a computer. How many times is the program iterated when the computational error reaches of the order of unity? Assume the program has ten digits of precision.

Chaos is quite vulnerable to numerical error, which is controlled by the precision of the computer. This amounts to introducing random noise in the tenth digit, for example, in an algebraic process with infinite precision. Chaos is very sensitive to this kind of small disturbance. This fact has a great implication in chaos as a natural phenomenon. This is because there always exists random disturbance in the natural world. This system of 'chaos +, small disturbance' makes chaos, which is essentially deterministic, indeterministic in practice. In other words, the system at any later time will be determined if a chaotic system, with a given map and an initial condition, has no disturbance. If, in contrast, there are disturbances, however small they may be, the value of the series can be specified only as an interval, such as [0, 1], or a probability distribution, at most, due to its sensitivity to the initial condition. This is an important practical aspect of chaos.

It is certainly required to introduce a more regorous definition of chaos than our intuitive one, namely 'an irregular oscillation generated by a simple rule', mentioned in this chapter. This amounts to defining what is meant by 'irregularity'. There are such irregularities as those series generated by maps previously mentioned, a series of pips of a die, or a sequence of faces of a coin when it is tossed up repeatedly.

Chaos is generated not only by a one-dimensional map but also by more complicated maps or a system of differential equations. It should also be noted that chaos is observed in real world phenomena. The existence of chaos in reality makes it not only a subject of mathematics or computational physics but also a subject of experiments or observations in many fields.

1.3 Chaos in Nature

Let us conclude this chapter by remarking the relation between one-dimensional maps and chaos in the real world. The significance of chaos is also mentioned.

One-dimensional maps, being so simple, seem to have nothing to do with existing chaos in the real world. This is not the case, however, and they may be extracted from irregular oscillations found in experiments or observations. The following example suffices to explain this.

Suppose the irregular oscillation shown in figure 1.14 is observed. These observational data are based on a signal in the *Belousov–Zhabotinsky reaction*.[4] The quickest way to extract chaotic behaviour from this oscillation is to define the series $A_1, A_2, A_3, \ldots, A_n, \ldots$ as the values of the maximum and the minimum of the wave. Then one plots the points (A_i, A_{i+1}) $(i = 1, 2, 3, \ldots)$ in a plane as

[4] A chemical reaction discovered by B Z Belousov and extended by A M Zhabotinsky. Bromine malonic acid ($BrCH(COOH)_2$) is produced by oxidizing malonic acid ($CH_2(COOH)_2$) in the reaction. The long term oscillation and spatial pattern formation with diffusion are observed in the concentration of ions, such as Br^- or Ce^{3+}, as the reaction proceeds.

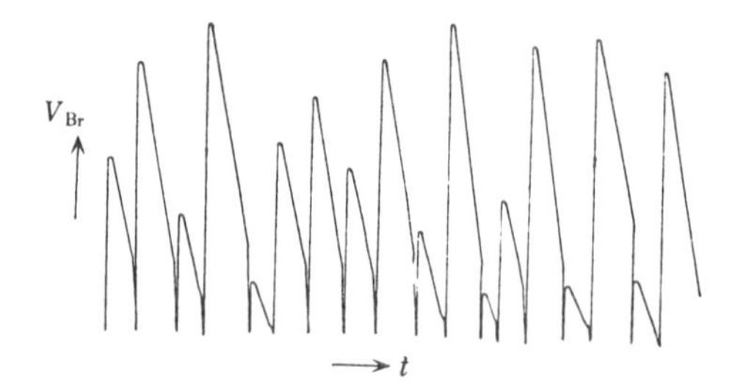

Figure 1.14. An irregular oscillation of the concentration of the bromine ion in the Belouzov–Zhabotinsky reaction (schematic).

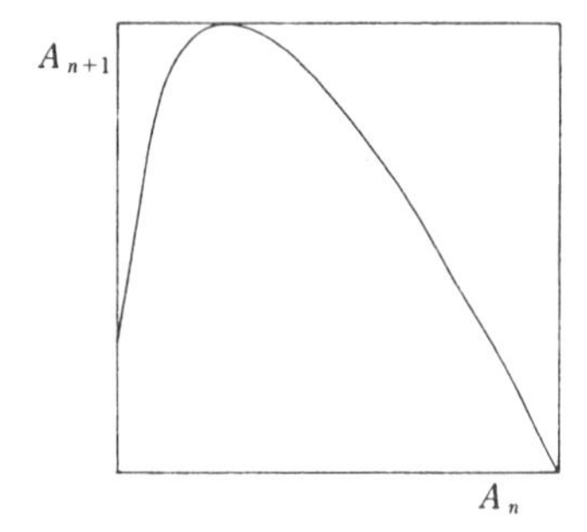

Figure 1.15. A curve obtained from the plot of the extrema (A_n, A_{n+1}) $(n = 1, 2, 3, \ldots)$.

shown in figure 1.15, from which one finds that this plot defines a single-valued map. If this map is expressed as $A_{n+1} = f(A_n)$, the peaks of the irregular wave of figure 1.14 are found to be governed by this simple rule. (Note that a one-dimensional map of this kind is not always deduced from an irregular wave. One should think a simple map may be obtained *if one is fortunate*.)

Let us consider why a one-dimensional map has been obtained by processing the data. This is because the physical system is trapped in the *strange attractor*. This attractor will be studied in detail in chapter 4.

A one-dimensional map thus emerges from a real system, which shows that it is fundamental as well as realistic.

What are the significances of chaos in the mechanical view of the world established so far? One of them is the expansion of the way we think of mechanics or, in a wider sense, of Nature. One finds no system in undergraduate (classical) mechanics which shows such a complex behaviour as chaos. For example, free fall, a simple pendulum and planetary motion have solutions with simple behaviour. In other words, these dynamical systems are mathematically integrable and their solutions are well behaved. It seems that these examples

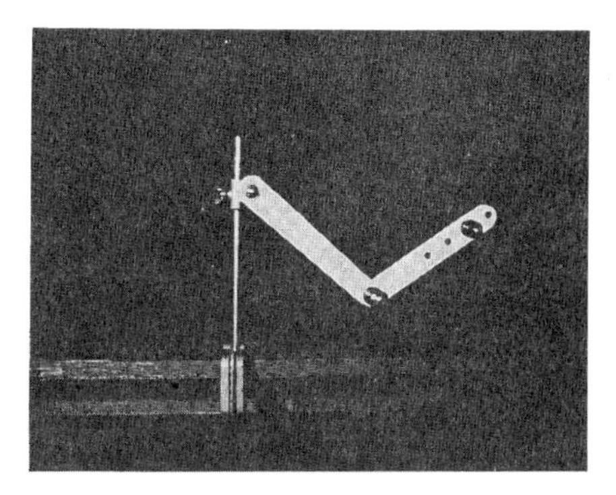

Figure 1.16. Chaos in a double pendulum, see appendix 4C.

have led to the illusion that any deterministic system should be equally well behaved. Chaos is a counterexample to this viewpoint in mechanics. It also forces us to think over the bias that a system with a complex behaviour is a many-body system and the system itself must be complex. Figure 1.16 shows a double pendulum, which is a simple system, yet showing a complex behaviour. This is an example of a conserved system considered in chapter 4. Although this is a familiar example, this is not a chaos associated with an attractor, as mainly considered in this book. Another significance of chaos is that the future behaviour of the combined system (chaos + small disturbances) is unpredictable because chaos shows an orbital instability. This *unpredictability* appears in a system for which macroscopic classical mechanics is applied and hence has practical importance, in contrast with the uncertainty principle in quantum mechanics, which is often discussed in the context of epistemology in a microscopic world.

It is quite recent that the existence of chaos in a realistic system has attracted much attention and it is still a new topic in physics. Although chaos is a unique phenomenon as yet, it will become a standard subject, such as a regular oscillation, and its position in Nature will be properly appreciated.

Although chaos generated by a one-dimensional map seems to be a toy, it serves as the foundation of chaos in general. Chapters 2 and 3 are devoted to the nature of chaos related to one-dimensional maps and also the criteria for being chaotic. Chaos is characterized not by its irregular behaviour in the series of the oscillations but in the following way:

(1) A condition given by Li and Yorke is 'if a map has periodic motions with the period 3, it leads to chaos', (2) related to (1) is the 'positivity of the topological entropy' and (3) 'positivity of the Lyapunov exponent given by the logarithm of the expansion rate of the map'. Although the following exposition

might seem to be slightly mathematical, it is written so that the reader will go through it with no difficulty if read in order.

Chapter 2

Li–Yorke chaos, topological entropy and Lyapunov number

The criteria for chaos will be stated in this chapter. The historical Li–Yorke theorem gives a criterion for the existence of nonperiodic orbits in a one-dimensional map. The topological entropy and the Lyapunov number describe the folding and stretching property of a chaotic map and are important in practice.

2.1 Li–Yorke theorem and Sharkovski's theorem

2.1.1 Li–Yorke theorem

The paper 'Period three implies chaos' by Li and Yorke, published in 1975 [1], had enomous impact in the research of chaos. The first half of the theorem states that

'if a continuous function f defined on the interval $[a, b]$ satisfies $a \leq f(x) \leq b$ and, moreover, f has period 3, then f has arbitrary periods'.

Here 'f has period 3' means that there exists a number c such that $c, f(c), f^2(c)$ $(= f(f(c)))$ are all different and $f^3(c)$ $(= f(f^2(c))) = c$ (see figure 2.1). Such c is called a *period 3 point*. More generally, a function f is said to have a period k point if there exists a point c such that the orbit of f starting from c, that is, $\{c, f(c), f^2(c), \ldots, f^{k-1}(c)\}$ comes back to c for the first time after k steps. In other words, f has a period k if there exists c such that points $c, f(c), f^2(c), \ldots, f^{k-1}(c)$ are all different but $f^k(c)$ $(= f(f^{k-1}(c))) = c$. We also define $f^0(c) = c$. In particular, a point c is called a fixed point of f if $c = f(c)$, since the orbit starting from c remains at c forever. This c is a crossing point of $y = f(x)$ and $y = x$ as shown in figure 2.2. Similarly a period k point is a crossing point of $y = f^k(c)$ and $y = x$. Note, however, that there are periodic points with smaller periods among these crossing points. For example, there are period 3 points among the solutions of $f^6(x) = x$ because $f^3(x) = x$ implies $f^6(x) = f^3(f^3(x)) = f^3(x) = x$ (see figure 2.3). In general, if p is a

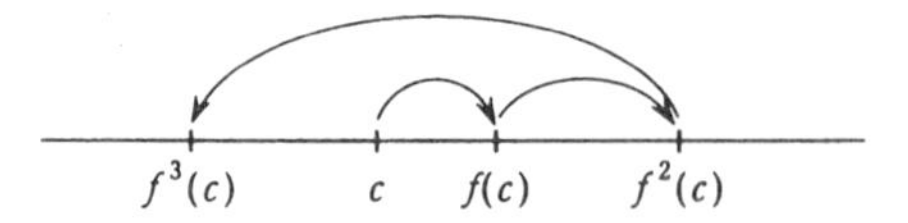

Figure 2.1. A point c is mapped to $f(c)$, $f^2(c)$, $f^3(c)$ by a function f. Here $f^3(c) < c$ and hence c is not a period 3 point. Actually, the condition in the Li–Yorke theorem is that 'there exists a point c such that $f^3(c) \leq c < f(c) < f^2(c)$ or $f^3(c) \geq c > f(c) > f^2(c)$'.

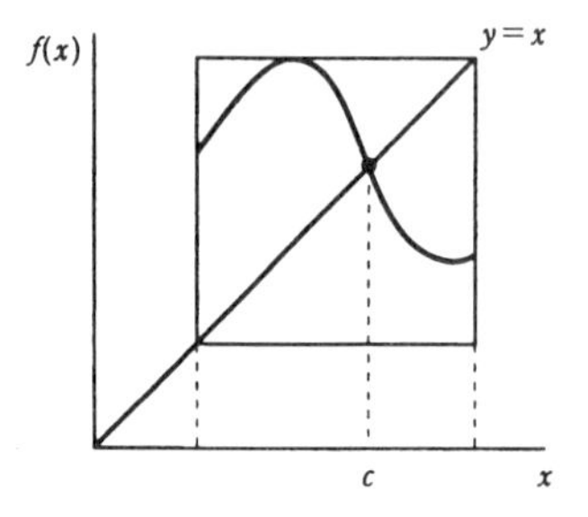

Figure 2.2. A fixed point $f(c) = c$ of a map f.

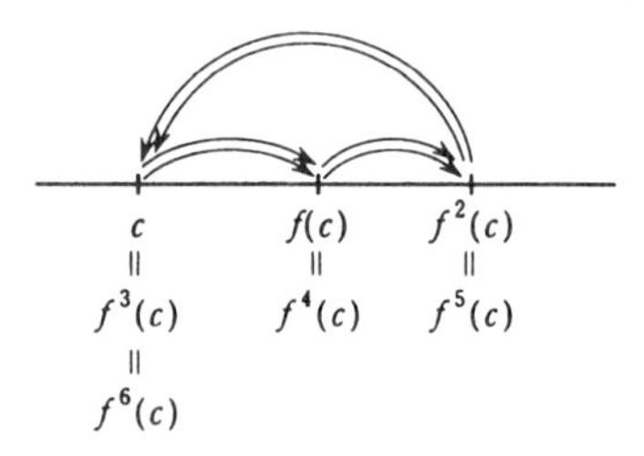

Figure 2.3. The orbit above is said to be of period 3 and not of period 6.

divisor of n, there are solutions of $f^p(x) = x$ among solutions of $f^n(x) = x$ since $f^p(x) = x$ implies $f^{pq}(x) = x$.

If we write $f(x_0) = x_1$, $f(x_1) = x_2, \ldots, f(x_{n-1}) = x_n$, then the relation

$$f^{n+m}(x_0) = f^n(f^m(x_0)) = f^m(f^n(x_0)) = x_{n+m}$$

is satisfied.

The Li–Yorke theorem is a truly amazing one since the condition that f has a period 3 necessarily implies f has arbitary periods.

Suppose a function f defined on $I = [a, b]$ takes its value $f(x)$ in the same interval I. Then we say 'f is a map from I to I', by using a word *map* (or *transformation*), which is more general than 'function'. We use these words since f moves a point in I iteratively to another point in I. The series $\{c, f(c), f^2(c), \ldots, f^n(c), \ldots\}$ is called an *orbit* of f with the *initial value* c. If, in particular, c is a period k point, this series is called a *period k orbit*. It may happen that, even though the first several terms of a series may not be periodic, the rest of the series forms a periodic orbit. Let us consider the tent map (1.3)

$$T(x) = \begin{cases} 2x & \left(0 \leq x < \dfrac{1}{2}\right) \\ 2 - 2x & \left(\dfrac{1}{2} \leq x \leq 1\right) \end{cases}$$

for example. The orbit with the initial value $\frac{1}{3}$ is $\{\frac{1}{3}, \frac{2}{3}, \frac{2}{3}, \ldots\}$ while one with the initial value $\frac{1}{10}$ is $\{\frac{1}{10}, \frac{1}{5}, \frac{2}{5}, \frac{4}{5}, \frac{2}{5}, \frac{4}{5}, \ldots\}$. The former (latter) has a period 1 (3), although $\frac{1}{3}, \frac{1}{10}, \frac{1}{5}$ are not periodic. These points are called *eventually periodic* points.

2.1.2 Sharkovski's theorem

The Li–Yorke theorem claims that a map f has periodic orbits of arbitrary periods provided that it has a period 3 orbit. It was A N Sharkovski, however, who found a more elaborate theorem in 1964 [2] prior to the work of Li and Yorke. His work was written in Russian in a Ukranian mathematical journal and had not attracted the attention of the Western mathematicians.

Suppose f is a continuous map from I to I. Let us write $n \Rightarrow m$ if the existence of a period n point of f necessarily implies that of a period m point. Then Sharkovski's theorem claims that

$$\begin{aligned} 3 \Rightarrow 5 \Rightarrow 7 \Rightarrow 9 \Rightarrow 11 \Rightarrow \ldots \Rightarrow 2n+1 \Rightarrow \ldots \Rightarrow \\ \Rightarrow 2 \cdot 3 \Rightarrow 2 \cdot 5 \Rightarrow 2 \cdot 7 \Rightarrow \ldots \Rightarrow 2(2n+1) \Rightarrow \ldots \Rightarrow \\ \Rightarrow 2^2 \cdot 3 \Rightarrow 2^2 \cdot 5 \Rightarrow \ldots \Rightarrow 2^2(2n+1) \Rightarrow \ldots \Rightarrow \\ \ldots \\ \Rightarrow 2^m \cdot 3 \Rightarrow 2^m \cdot 5 \Rightarrow \ldots \Rightarrow 2^m(2n+1) \Rightarrow \ldots \Rightarrow \\ \ldots \\ \Rightarrow \ldots \Rightarrow 2^k \Rightarrow 2^{k-1} \Rightarrow \ldots \Rightarrow 16 \Rightarrow 8 \Rightarrow 4 \Rightarrow 2 \Rightarrow 1. \end{aligned}$$

Figure 2.4 explains why period 3 implies period 5 and period 7. The sequence above defines an *ordering*, in a sense, in the set of natural numbers since all of them appear once and only once there. If this order is understood as an order of 'strength', the sequence states that 3 is the strongest one, odd numbers (except 1) are stronger than even numbers and the powers of 2 and 1 ($= 2^0$) are the weakest. The proof of the theorem can be found in [3].

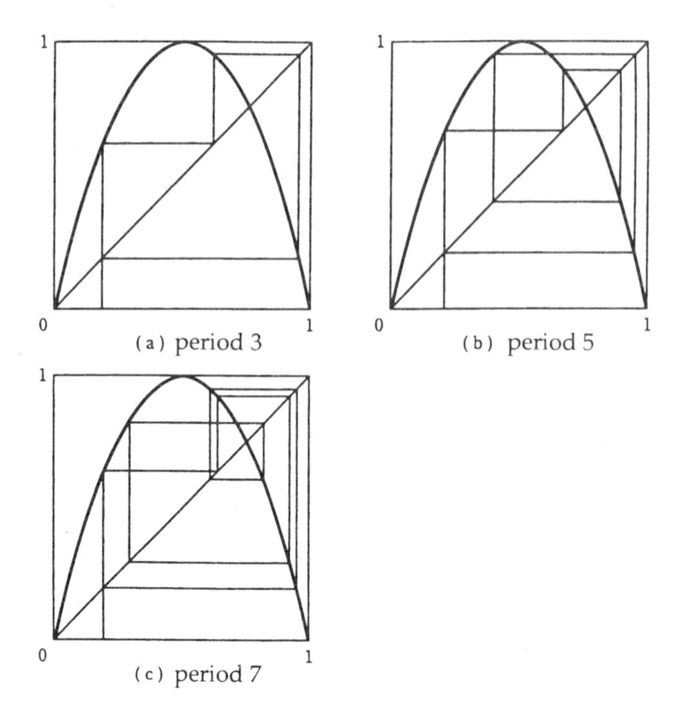

Figure 2.4. An example of 'period 3 implies period 5 and period 7'.

2.2 Periodic orbits

2.2.1 Number of periodic orbits

Let us consider, as a concrete example of a map from $I = [0, 1]$ to I, the tent map $T(x)$ introduced in chapter 1, the logistic map

$$L(x) = 4x(1 - x)$$

and the binary transformation[1]

$$B(x) = \begin{cases} 2x & \left(0 \leq x < \dfrac{1}{2}\right) \\ 2x - 1 & \left(\dfrac{1}{2} \leq x < 1\right) \end{cases}$$

and study their periodic orbits and their number.

[1] Given a number x, one keeps only the decimal places of $2x$ in the original binary transformation. The domain and the range are thus taken to be $I = [0, 1)$.

The period 1 points of $T(x)$ are the intersections of $y = T(x)$ and $y = x$, namely two points $x = 0, \frac{2}{3}$. The period 2 points are the intersections of $y = T^2(x)$ and $y = x$, namely two points $x = \frac{2}{5}, \frac{4}{5}$, where other two intersections $x = 0$ and $\frac{2}{3}$ are omitted since they are the period 1 points. The period 3 points are the intersections of $y = T^3(x)$ and $y = x$, i.e., six points $x = \frac{2}{9}, \frac{2}{7}, \frac{4}{9}, \frac{4}{7}, \frac{6}{7}, \frac{8}{9}$, where the points $x = 0$ and $\frac{2}{3}$ are again omitted. See figure 2.5(a). In general, the period n points are the rest of the 2^n intersections of $y = T^n(x)$ and $y = x$ with all the period p points subtracted, where p divides n. The number of the period n points, denoted by $A(n)$, is given by

$$A(n) = \sum_{p|n} \mu\left(\frac{n}{p}\right) 2^p \tag{2.1}$$

where the Möbius inversion formula has been used (see appendix 2A). In the equation above, the symbol $p|n$ means that p divides n and $\sum_{p|n}$ means the summation over all such p should be taken. The function $\mu(x)$, called the Möbius function, is defined on the set of the natural numbers and takes only three values ± 1 and 0. The factor 2^p is the number of intersections of $y = T^p(x)$ and $y = x$.

The first several $A(n)$ are

$$A(1) = A(2) = 2, \;\; A(3) = 6, \;\; A(4) = 12, \;\; A(5) = 30, \ldots$$

while $A(n)$ for larger n are

$$A(10) = 990, \;\; A(20) = 1047\,540, \;\; A(50) = 1125\,899\,873\,287\,200, \ldots.$$

If the values just above are compared to

$$2^{10} = 1024, \;\; 2^{20} = 1048\,576, \;\; 2^{50} = 1125\,899\,906\,842\,624, \ldots$$

one finds that the ratio of $A(n)$ and 2^n approaches unity as $n \to \infty$.

Since only the fact that the number of the intersections of $y = T^n(x)$ and $y = x$ is 2^n is used to obtain $A(n)$, one finds that the number of period n points for $L(x)$ is again given by $A(n)$ since the number of intersections of $y = L^n(x)$ with $y = x$ is also 2^n (see figure 2.5(b)).

Let us next consider the binary transformation $B(x)$, which is defined on $0 \le x < 1$. Since $x = 1$ is omitted from the domain, the number of intersections of $y = B^n(x)$ and $y = x$ is $2^n - 1$, instead of 2^n (see figure 2.5(c)). In this case, the number of period n points is given, similarly to $A(n)$, by

$$\sum_{p|n} \mu\left(\frac{n}{p}\right)(2^p - 1) = \sum_{p|n} \mu\left(\frac{n}{p}\right) 2^p - \sum_{p|n} \mu\left(\frac{n}{p}\right).$$

This expression is equal to $A(n)$ for $n \ge 2$ since $\sum_{p|n} \mu\left(\frac{n}{p}\right) = 0$ in this case. Thus the number of the period n points is common to $T(x)$, $L(x)$ and $B(x)$ for $n \ge 2$.

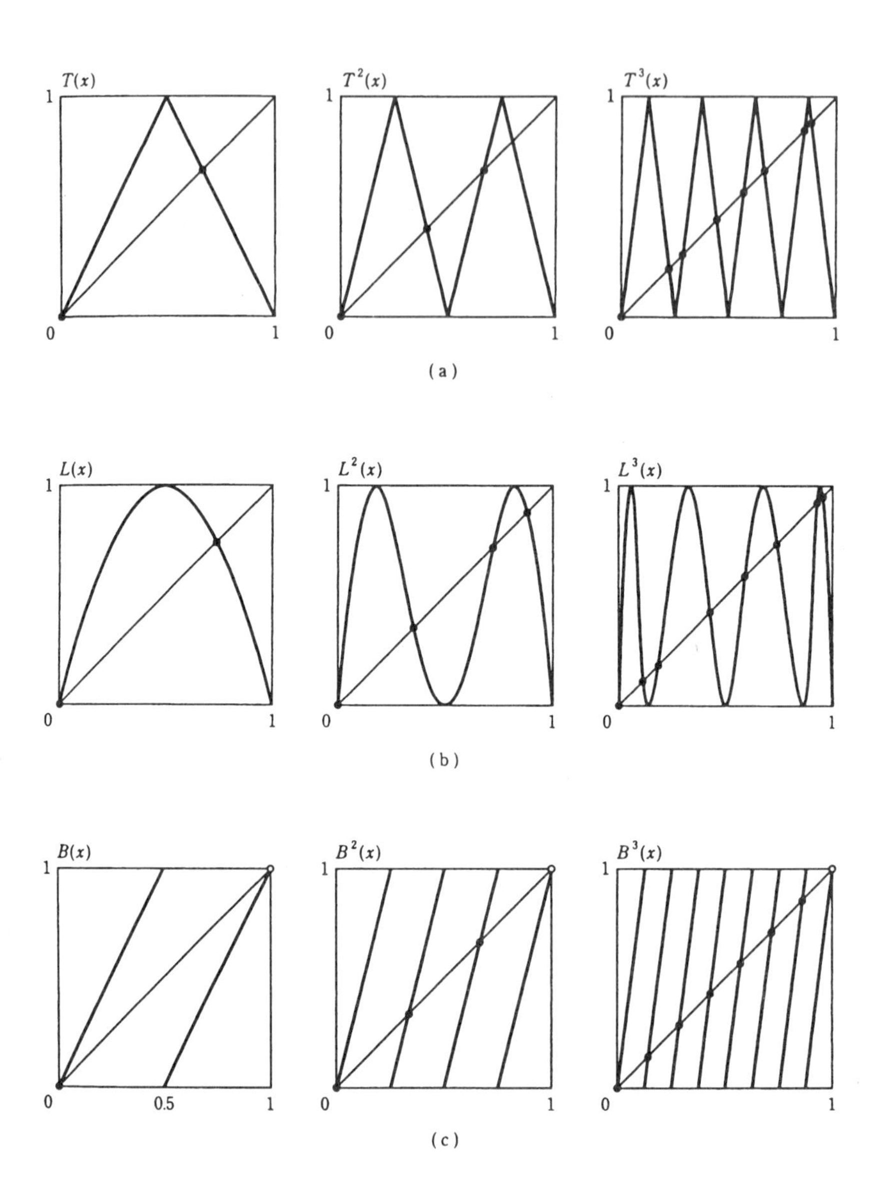

Figure 2.5. (a) The intersections of $T(x)$, $T^2(x)$ and $T^3(x)$ with $y = x$. (b) The intersections of $L(x)$, $L^2(x)$ and $L^3(x)$ with $y = x$. (c) The intersections of $B(x)$, $B^2(x)$ and $B^3(x)$ with $y = x$.

Problem 1. Show that the number of period n orbits of $T(x)$, $L(x)$ and $B(x)$ is $A(n)/n$.

The characteristic of the binary transformation $B(x)$ becomes clearer if x is expressed in binary fractions as

$$x = (0.x_1x_2\ldots x_n\ldots)_2 = \frac{x_1}{2} + \frac{x_2}{2^2} + \ldots + \frac{x_n}{2^n} + \ldots \qquad (x_n = 0, 1).$$

Similarly to the expression for ordinary decimal numbers such as

$$0.1 = 0.099\,999\ldots \qquad 0.12 = 0.119\,9999\ldots$$

it is possible to express a finite binary fraction in two ways. In the following, we employ a convention in which finite fractions such as

$$\frac{3}{4} = \frac{1}{2} + \frac{1}{2^2} = (0.11)_2 \; (= (0.101\,1111\ldots)_2)$$

$$\frac{5}{8} = \frac{1}{2} + \frac{0}{2^2} + \frac{1}{2^3} = (0.101)_2 \; (= (1.100\,111\,11\ldots)_2)$$

are represented as finite fractions with infinite zeros, as $(0.110\,0000\ldots)_2$ or $(0.101\,0000\ldots)_2$, unless otherwise stated.

The following facts are true for binary fractions, similarly to decimal fractions.

(i) x is a finite binary number if and only if x is of a rational number of the form $j/2^k$ (such as $3/2^2 = 1/2 + 1/2^2$).

(ii) Any number x ($\neq k/2^j$) in the interval $[0, 1]$ is uniquely expressed as an infinite binary fraction. A rational number is then expressed as a recurring fraction while an irrational number as a nonrecurring infinite fraction.

It follows from a property of the binary transformation $B(x)$ that $x < 1/2$ provided that $x_1 = 0$ in $x = (0.x_1x_2\ldots x_n\ldots)_2$. One then finds

$$\begin{aligned} B(x) = 2x &= 2\left(\frac{x_2}{2^2} + \frac{x_3}{2^3} + \ldots + \frac{x_n}{2^n} + \ldots\right) \\ &= \frac{x_2}{2} + \frac{x_3}{2^2} + \ldots + \frac{x_{n+1}}{2^n} + \ldots = (0.x_2x_3\ldots x_{n+1}\ldots)_2. \end{aligned}$$

On the other hand, one has $\frac{1}{2} \leq x < 1$ if $x_1 = 1$ and hence

$$\begin{aligned} B(x) = 2x - 1 &= 2\left(\frac{1}{2} + \frac{x_2}{2^2} + \ldots + \frac{x_n}{2^n} + \ldots\right) - 1 \\ &= \frac{x_2}{2} + \frac{x_3}{2^2} + \ldots + \frac{x_{n+1}}{2^n} + \ldots = (0.x_2x_3\ldots x_{n+1}\ldots)_2. \end{aligned}$$

Therefore, the binary transformation shifts each digit of $x = (0.x_1x_2\ldots x_n\ldots)_2$ by one towards the left. Accordingly it follows that

$$B^2(x) = (0.x_3x_4\ldots x_{n+2}\ldots)_2, \quad \ldots, \quad B^n(x) = (0.x_{n+1}x_{n+2}\ldots x_{2n}\ldots)_2.$$

One easily finds that the periodicity of the orbits is related to binary recurring fractions if one notices that the binary transformation may be regarded as a *shift transformation* of a binary number towards the left. For example,

$$(0.110\,111\,01\ldots)_2 = (0.\dot{1}10\dot{1})_2 = \frac{13}{15}$$

is a recurring fraction with the recurring unit of the length 4 and hence it is a period 4 point of $B(x)$.

Problem 2. Show that $(0.\dot{1}10\dot{1})_2 = \frac{13}{15}$ and find an orbit of $B(x)$ with period 4 starting from this point.

2.2.2 Stability of orbits

Let us consider a transformation

$$L_R(x) = Rx(1-x)$$

defined on an interval $I = [0, 1]$, which generalizes the logistic map $L(x) = 4x(1-x)$. Here the parameter R is restricted within the range $0 < R \leq 4$. It follows that $0 \leq L_R(x) \leq 1$ for any $0 \leq x \leq 1$ if and only if R takes this range. This transformation L_R will be analysed in chapter 3 in detail. We take $0 < R < 3$ for the time being.

Let us consider two orbits $\{x, L_R(x), L_R^2(x), \ldots\}$ and $\{y, L_R(y), L_R^2(y), \ldots\}$ with $x \neq y$, none of which is equal to any of 0, $1 - 1/R$ and 1. Since

$$\lim_{n\to\infty} L_R^n(x) = \lim_{n\to\infty} L_R^n(y) = 1 - \frac{1}{R}$$

as seen from figure 2.6(c), one has

$$\lim_{n\to\infty} |L_R^n(x) - L_R^n(y)| = 0.$$

In general, if two orbits $\{x, f(x), f^2(x), \ldots\}$ and $\{y, f(y), f^2(y), \ldots\}$ satisfy

$$\lim_{n\to\infty} |f^n(x) - f^n(y)| = 0$$

they are said to *approach asymptotically*. This includes the case where $f^n(x) = f^n(y)$ for any $n > n_0$. Let x be a fixed point of a transformation f, namely $f(x) = x$. If an orbit $\{y, f(y), f^2(y), \ldots\}$, whose initial point y ($\neq x$) is an arbitary point in a neighbourhood[2] of x, satisfies

$$\lim_{n\to\infty} |f^n(y) - x| = 0$$

[2] y is a point whose distance from x is less than a positive number ε. The set of such points is called the *ε-neighbourhood*. For a real line this means an open interval $(x - \varepsilon, x + \varepsilon)$ and for a plane this is inside the circle with radius ε centred at x.

then x is said to be *stable*. This also implies that two orbits of f with the initial points x and y, respectively, approach asymptotically. For example, the point 0 is a stable fixed point of L_R for $0 < R \leq 1$. For $1 < R$, in contrast, 0 is no longer stable but an unstable point. In fact, one obtains $\lim_{n\to\infty} L_R^n(y) = 1 - 1/R$ for $1 < R < 3$ even if one started with a point y arbitrarily close to 0 as shown in figure 2.6(c). It turns out that the point $1 - 1/R$ is the stable fixed point of L_R in the present case.

Since $f^k(x) = x$, if x is a period k point of f, x is a fixed point of f^k. If, furthermore, x is stable, it is called a *stable period k point* of f.

The stability of a fixed point x of f is determined from whether $|f'(x)| < 1$ or $|f'(x)| > 1$ (see figures 2.6(a) and (b)). For example one finds $|L'_R(1-1/R)| = |2 - R| < 1$ for $1 < R < 3$.

Comments on the derivative of $f^k(x) = f(f^{k-1}(x))$ and related problems are in order before we close this section. One simply uses the chain rule to

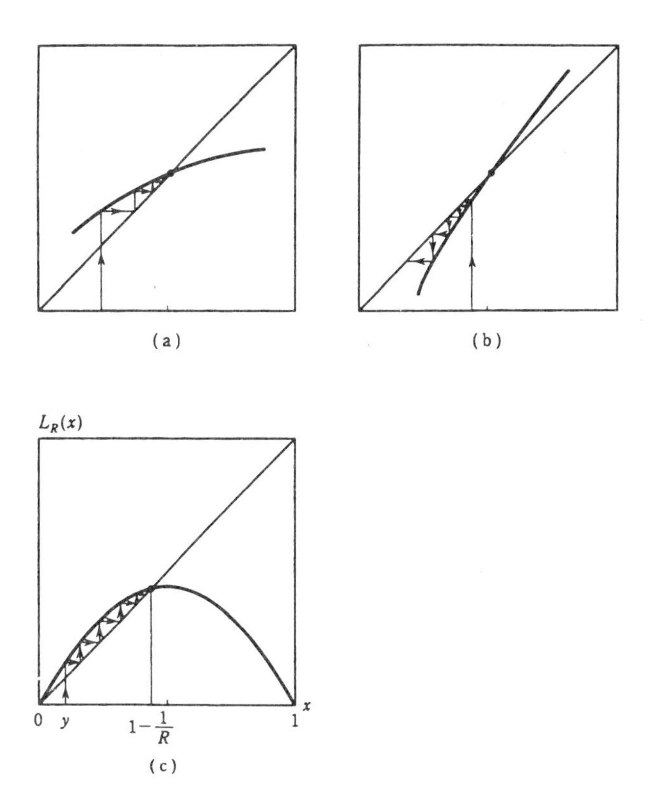

Figure 2.6. (a) A stable fixed point. (b) An unstable fixed point. (c) The stable fixed point of $L_R(x) = Rx(1-x)$ $(1 < R < 3)$.

obtain

$$\frac{d}{dx}f^2(x) = \frac{d}{dx}f(f(x)) = f'(f(x))f'(x) = \frac{df^2}{df}\frac{df}{dx}$$

$$\frac{d}{dx}f^3(x) = f'(f^2(x))\frac{d}{dx}f^2(x) = f'(f^2(x))f'(f(x))f'(x) = \frac{df^3}{df^2}\frac{df^2}{df}\frac{df}{dx}.$$

By repeating this, one obtains

$$\frac{d}{dx}f^n(x) = f'(f^{n-1}(x))f'(f^{n-2}(x))\dots f'(f(x))f'(x)$$
$$= \frac{df^n}{df^{n-1}}\frac{df^{n-1}}{df^{n-2}}\dots\frac{df^2}{df}\frac{df}{dx}. \tag{2.2}$$

Problem 3. Let x_0 be a period n point of a transformation f and let $x_1 = f(x_0), x_2 = f(x_1) = f^2(x_0), \dots, x_{n-1} = f^{n-1}(x_0)$. Show that

$$g'(x_0) = f'(x_0)f'(x_1)f'(x_2)\dots f'(x_{n-1})$$

where g denotes the transformation f^n. Show, from this, that

$$g'(x_0) = g'(x_1) = \dots = g'(x_{n-1}) = f'(x_0)f'(x_1)\dots f'(x_{n-1}).$$

(Accordingly all the points on a single period n orbit are stable or unstable *simultaneously*.)

Problem 4. Let $f(x_0) = x_0$, $f'(x_0) = -1$ and $g(x) = f^2(x)$. Show that $g'(x_0) = 1$ and $g''(x_0) = 0$.

2.3 Li–Yorke theorem (continued)

The first half of the Li–Yorke theorem was explained in 2.1.1. The last half of the theorem states, under the same condition, that 'a map f from the interval I to I has period 3'; the following conclusions are derived.

There is an uncountable set S (see appendix 2B) of nonperiodic points of f satisfying (i) and (ii) below.

(i) An arbitrary point x of S and any periodic point p of f satisfy

$$\overline{\lim_{n\to\infty}}|f^n(x) - f^n(p)| > 0. \tag{2.3}$$

(ii) Arbitrary points x and y of S satisfy

$$\overline{\lim_{n\to\infty}}|f^n(x) - f^n(y)| > 0 \quad \underline{\lim_{n\to\infty}}|f^n(x) - f^n(y)| = 0. \tag{2.4}$$

The set S is called the *scrambled set*. Some call S a scrambled set even when it is not uncountable. The symbols $\overline{\lim}_{n\to\infty}$ and $\underline{\lim}_{n\to\infty}$ generalize $\lim_{n\to\infty}$ and are explained in appendix 2C.

The condition (i) above requires that an orbit starting from a point in S does not approach asymptotically any periodic orbit, that is, it cannot satsify $\lim_{n\to\infty} |f^n(x) - f^n(p)| = 0$. The condition (ii) states that orbits starting from two different points in S do not approach each other asymptotically but they can be arbitrarily close to each other (i.e., $\underline{\lim}_{n\to\infty}|f^n(x) - f^n(y)| = 0$) on the way. Moreover there must be very many (uncountable!) such initial points. It was thought that these conditions were relevant for the explanation of chaos.

We say *a map f is chaotic in the sense of Li–Yorke (or simply f is a Li–Yorke chaos) if f has an uncountable scrambled set.* The condition 'f has the period 3' in the Li–Yorke theorem is a sufficient condition for f to have Li–Yorke chaos. As a matter of fact,

'if f has a period $2^n(2m+1)$ $(n \geq 0, m \geq 1)$, then f is a Li–Yorke chaos (period $\neq 2^n$ implies chaos). That is, there exists an uncountable scrambled set S such that equations (2.3) and (2.4) are true'.[3]

2.4 Scrambled set and observability of Li–Yorke chaos

2.4.1 Nathanson's example

Consider a map $f(x)$ from an interval $[0, 1]$ to $[0, 1]$ defined by (M Nathanson [5])

$$f(x) = \begin{cases} x + \dfrac{1}{p} & \left(0 \leq x \leq 1 - \dfrac{1}{p}\right) \\ 1 - \dfrac{1-\delta}{\delta}\left(x - \dfrac{p-1}{p}\right) & \left(1 - \dfrac{1}{p} \leq x \leq 1 - \dfrac{1}{p} + \delta\right) \\ x - \dfrac{p-1}{p} & \left(1 - \dfrac{1}{p} + \delta \leq x \leq 1\right) \end{cases} \tag{2.5}$$

(see figure 2.7). Here $p \geq 3$ is an integer and $0 < \delta < 1/2^p$. Then f satisfies

$$f^3\left(1 - \frac{2}{p}\right) \leq 1 - \frac{2}{p} < f\left(1 - \frac{2}{p}\right) < f^2\left(1 - \frac{2}{p}\right)$$

(see the caption of figure 2.1) and hence satisfies the conditions in the Li–Yorke theorem. In fact,

$$f\left(1 - \frac{2}{p}\right) = 1 - \frac{1}{p} \qquad f\left(1 - \frac{1}{p}\right) = 1 \qquad f(1) = \frac{1}{p}$$

[3] It is known that there exists a map with infinite periods of the form 2^n but no period of the form $2^n(2m+1)$, which is still a Li–Yorke chaos [4].

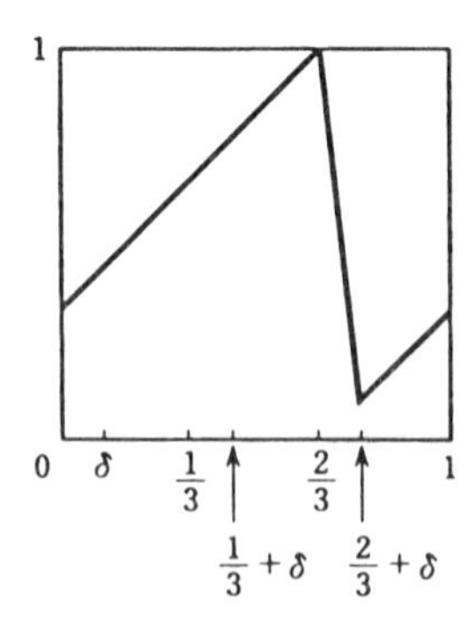

Figure 2.7. Nathanson's example $\left(p = 3, \delta = \frac{1}{8}\right)$.

and the condition $\frac{1}{p} \le 1 - \frac{2}{p}$ is equivalent to $3 \le p$. Therefore f is a Li–Yorke chaos although the scrambled set S associated with f is a set with zero measure.[4] In fact, in the example shown in figure 2.7, the inverval $[\delta, \frac{1}{3}]$ is parallel transported to $[\frac{1}{3} + \delta, \frac{2}{3}]$, $[\frac{1}{3} + \delta, \frac{2}{3}]$ to $[\frac{2}{3} + \delta, 1]$ and $[\frac{2}{3} + \delta, 1]$ to $[\delta, \frac{1}{3}]$. Accordingly all the points in these three intervals are period 3 points. Moreover, one can prove that if a null set E is subtracted from the rest of the interval $[0, 1] - [\delta, \frac{1}{3}] - [\frac{1}{3} + \delta, \frac{2}{3}] - [\frac{2}{3} + \delta, 1]$, all the points x in the resulting set are eventually period 3 points. That is, there is a null set E such that, if it is subtracted from the interval $[0, 1]$, all the points in this set are either period 3 points or eventually period 3 points. According to the definition of the scrambled set S, however, periodic points and eventually periodic points are not contained in S. Therefore S must be a subset of E, which implies that S itself is a null set.

There are variety of choices of the scrambled set S given a transformation f. An uncountable subset S' of S is evidently a scrambled set for example. In the example above, however, any S is a null set.

2.4.2 Observability of Li–Yorke chaos

Such objects as lines in a plane may not be observable as a two-dimensional set since they have length only with no width and hence they have vanishing area. Therefore, if all the scrambled sets are null sets, as that of the Nathanson transformation, they are not observable, that is, the probability of choosing an initial condition in S is zero. Accordingly one might want to define that chaos in a transformation f is *observable* if f has a scrambled set with a positive measure ($\mu(S) > 0$). Here the following definition covering more general cases (where S is a nonmeasurable set, see appendix 2D) is employed [6, 7],

[4] See appendix 2D. A set with zero measure is also called a *null set*.

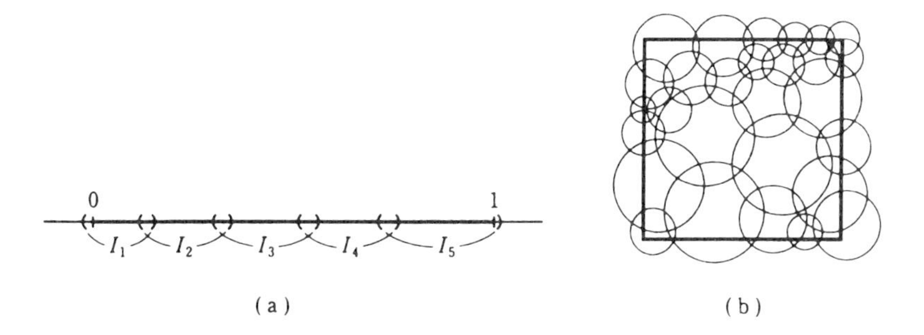

Figure 2.8. (a) An interval [0, 1] is covered with open intervals. (b) A square is covered with open discs.

'f has observable chaos if the inner measure (measured from inside) of the scrambled set S is positive ($\mu_*(S) > 0$)'.

Then the Nathanson transformation clearly has *unobservable* chaos, since the inner measure of a null set is zero. As for the tent map $T(x)$, it has been shown (J Smítal [8]) that there exists a scrambled set S whose outer measure (measured from outside) $\mu^*(S)$ is unity and unmeasurable (i.e., $\mu^*(S) > \mu_*(S)$). It is also shown for the tent map that any measurable scrambled set is a null set and hence the inner measure $\mu_*(S)$ of any scrambled set S is zero and, accordingly, chaos in $T(x)$ is unobservable. This is because $\mu_*(S)$ is the infimum[5] of the measure $\mu(S')$ of a measurable scrambled set S' contained in S ($S' \subset S$), namely $\mu_*(S) = \inf \mu(S')$, and $\mu(S') = 0$. It is known that this result for the tent map is true for more general cases. This fact indicates that the Li–Yorke theorem may shed light on a limited aspect of chaos. That is, chaos in the tent map or the logistic map, which is conjugate to the tent map, see 2.6.2, is observable in a computer, although the scrambled sets of these maps have zero Lebesgue measure and are hence unobservable. Therefore chaos we observe is not related to the scrambled set. This observable chaos will be mainly treated in this book in the following chapters.

2.5 Topological entropy

It is the topological entropy that is closely related to Li–Yorke chaos. Here the topological entropy $h(f)$ is defined for a map f from the interval $I = [a, b]$ to I, although more general definition may be possible. Open sets are introduced in the following since they are necessary for theoretical exposition. Here they simply mean open intervals or finite or infinite union thereof. For example, for

[5] Infimum is a generalization of minimum, see the footnote of appendix 2D. See books on the measure theory and the theory of the Lebesgue integral for this property of an inner measure.

the sets in a plane, they are open discs (inside a circle without the boundary) or unions of them.

When a figure is covered with finite (or infinite) open sets so that there are no openings, such a class of open sets is called an open cover (see figures 2.8(a) and (b)). Suppose open sets $I_1, I_2, \ldots, I_n, \ldots$ cover an interval I. Let $\alpha = \{I_n\}_{n\geq 1}$ denote the corresponding open covering. It may happen that not all of I_n are required to cover the interval I and it may be covered more efficiently without using all of the members of $\{I_n\}_{n\geq 1}$. Let $N(\alpha)$ be the least number of open sets in the subset of $\{I_n\}_{n\geq 1}$ covering the interval I. (Even when one started with an infinite number of open sets, one always finds a finite open covering.) Given a map f and a set E, the *inverse image* of E by f is the set of points in I that are mapped to E under f and denoted by $f^{-1}E$ $(= \{x \mid f(x) \in E\})$. The set $f^{-1}E$ is an open set when f is continuous and E is an open set. In the case of the tent map T and $E = \left(0, \frac{1}{2}\right)$, for example, $T^{-1}E$ is a union $\left(0, \frac{1}{4}\right) \cup \left(\frac{3}{4}, 1\right)$ of two intervals $\left(0, \frac{1}{4}\right)$ and $\left(\frac{3}{4}, 1\right)$ (see figure 2.9).

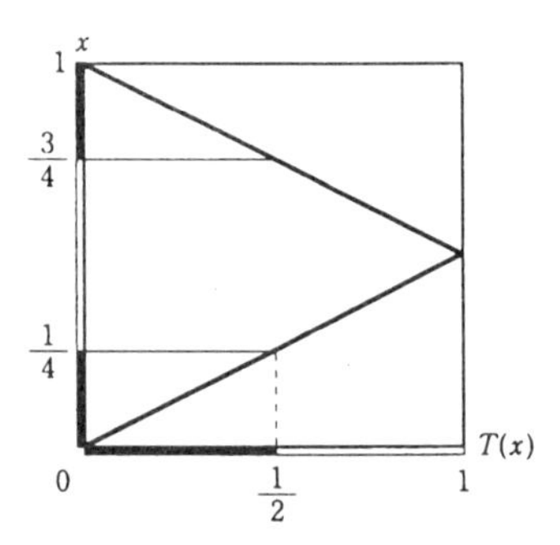

Figure 2.9. The inverse image of the tent map. The union of the black intervals (the white interval) on the x-axis (the ordinate) is the inverse image of the black (white) interval on the $T(x)$-axis (the abscissa).

It can be shown for a covering $\alpha = \{I_n\}_{n\geq 1}$ that the set $\{f^{-1}I_n\}_{n\geq 1}$ of the inverse image $f^{-1}I_n$ of each I_n with respect to f is again an open covering of I, which is denoted by $f^{-1}\alpha$. Let $\beta = \{J_m\}_{m\geq 1}$ be another open covering of I. Then the intersection $I_n \cap J_m$ of I_n and J_m for any n and m is an open set and the union of these open sets $\bigcup_{n\geq 1} \bigcup_{m\geq 1} (I_n \cap J_m)$ covers I. Thus $\{I_n \cap J_m\}$ is an open covering of I, which will be denoted as $\alpha \vee \beta$. This covering of I is clearly finer than α or β. Take $I = [0, 1]$ and open coverings $\alpha = \left\{\left[0, \frac{1}{2}\right), \left(\frac{1}{4}, 1\right]\right\}$ and $\beta = \left\{\left[0, \frac{3}{8}\right), \left(\frac{1}{4}, \frac{3}{4}\right), \left(\frac{5}{8}, 1\right]\right\}$[6] for example. Then $\alpha \vee \beta$ is the set of the intersections of the open intervals, given by $\left\{\left[0, \frac{3}{8}\right), \left(\frac{1}{4}, \frac{3}{8}\right), \left(\frac{1}{4}, \frac{1}{2}\right), \left(\frac{1}{4}, \frac{3}{4}\right), \left(\frac{5}{8}, 1\right]\right\}$. However, the interval [0, 1] is covered

[6] Semi-open intervals such as $\left[0, \frac{3}{8}\right)$ or $\left(\frac{1}{4}, 1\right]$ are employed when they contains the point 0 or 1, since the outside of the interval [0, 1] is not considered here.

with only three intervals, $\left[0, \frac{3}{8}\right), \left(\frac{1}{4}, \frac{3}{4}\right)$ and $\left(\frac{5}{8}, 1\right]$, and one obtains $N(\alpha \vee \beta) = 3$ as the smallest number of covering intervals.

Problem 5. Suppose α and β are open coverings of I and f is a continuous map from I to I. Show that $f^{-1}\alpha$ and $\alpha \vee \beta$ are also open coverings of I.

In particular, $\alpha \vee \beta$ is an open covering when $\beta = f^{-1}\alpha$. By adding to this an open covering $f^{-1}(f^{-1}\alpha)$, which will be written as $f^{-2}\alpha$, one obtains $(\alpha \vee f^{-1}\alpha) \vee f^{-2}\alpha = \alpha \vee f^{-1}\alpha \vee f^{-2}\alpha$, which is again an open covering. If this procedure is repeated $(n-1)$ times, one obtains an open covering $\alpha \vee f^{-1}\alpha \vee f^{-2}\alpha \vee \ldots \vee f^{-(n-1)}\alpha$, which is denoted as $\bigvee_{i=0}^{n-1} f^{-i}\alpha$. The number $N\left(\bigvee_{i=0}^{n-1} f^{-i}\alpha\right)$ is the smallest number of open sets in I that is necessary to cover I. In other words, this is the number of open sets in the most *efficient* open cover of I. The number $\log N\left(\bigvee_{i=0}^{n-1} f^{-i}\alpha\right)$ should be called, then, the entropy[7] of this open cover. The entropy

$$\frac{1}{n} \log N \left(\bigvee_{i=0}^{n-1} f^{-i}\alpha \right)$$

for each step approaches a limiting value as $n \to \infty$, which is denoted by $h(\alpha, f)$:

$$h(\alpha, f) = \lim_{n\to\infty} \frac{1}{n} \log N \left(\bigvee_{i=0}^{n-1} f^{-i}\alpha \right). \tag{2.6}$$

This value being dependent on the initial cover α, its supremum $\sup h(\alpha, f)$ for all the covers of I depends only on f. This quantity $h(f)$ is called the *topological entropy* of f:

$$h(f) = \sup h(\alpha, f). \tag{2.7}$$

The term 'topological' is named after a field in mathematics called topology, n which continuous maps and open sets are the objects of study.

The topological entropy $h(f)$ measures, in a sense, the complexity of the nap f. Figure 2.10 shows how the tent map T and its scalar multiple $\frac{2}{3}T$ map a losed cover[8] $\alpha = \left\{\left[0, \frac{1}{2}\right], \left[\frac{1}{2}, 1\right]\right\}$ of [0, 1]. Two maps are applied once, twice, hree times and four times and the figures show how the inverse images of α ehaves. The intersection of the inverse image at each stage is $\bigvee_{i=0}^{4} T^{-i}(\alpha)$ or $\bigvee_{i=0}^{4} \left(\frac{2}{3}T\right)^{-i}(\alpha)$ in respective case. One may see at this stage that T is more omplex than $\frac{2}{3}T$, that is, the former scrambles the points in [0, 1] more than ie latter. This fact is reflected upon the topological entropies $h(T)$ and $h(\frac{2}{3}T)$.

The entropy of a set $\{x_1, x_2, \ldots, x_N\}$ with N elements is defined as $\log_2 N$. This measures the omplexity of the set. Henceforth the base of logarithms is taken to be 2.

This cover is made of closed sets for the convenience of drawing. This should not cause any oblem in the description of the concept.

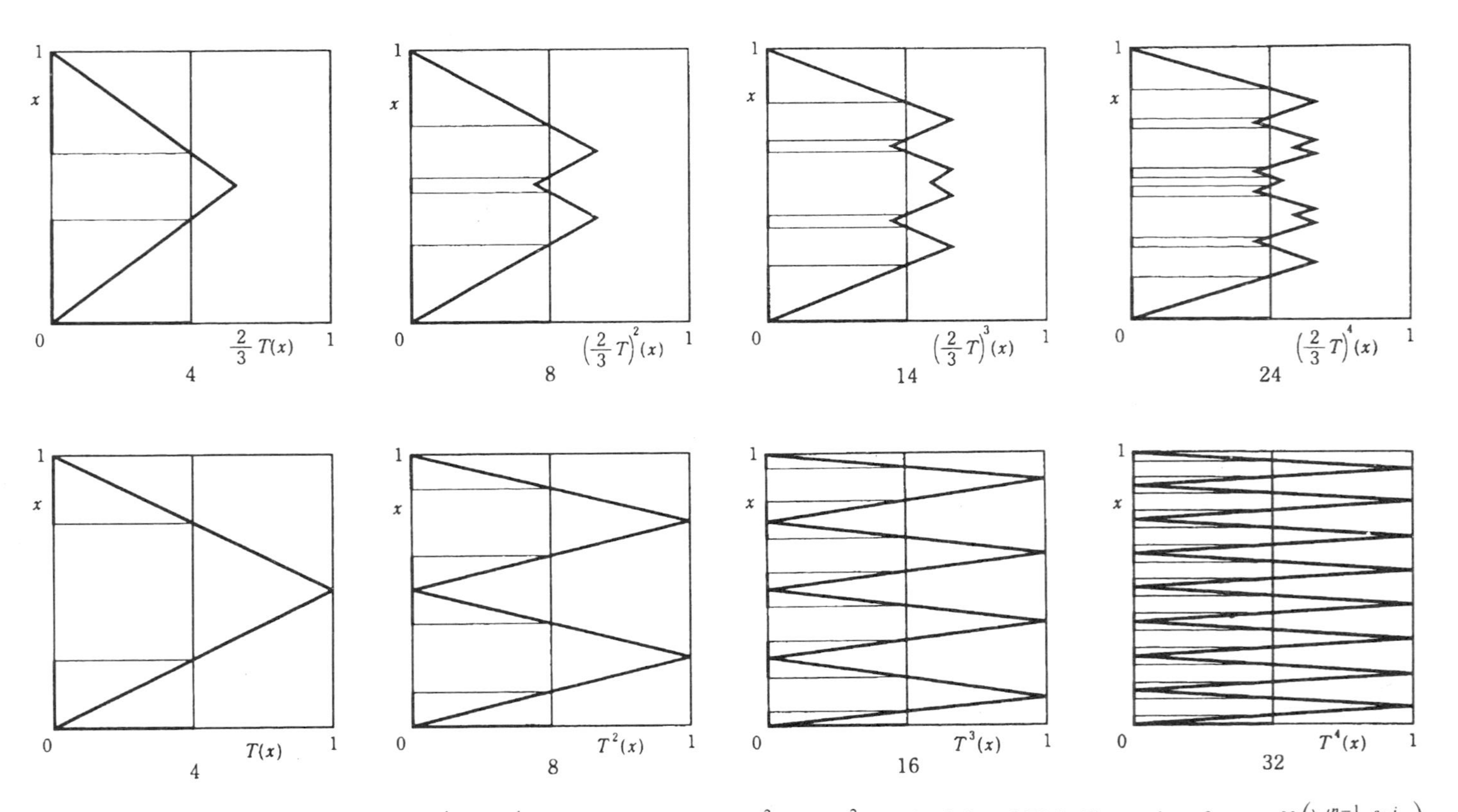

Figure 2.10. The inverse images of the sets $[0, \frac{1}{2}]$ and $[\frac{1}{2}, 1]$ under iterated maps of $\frac{2}{3}T(x) = \frac{2}{3}(1 - |1 - 2x|)$ and $T(x)$. The number of covers $N\left(\bigvee_{i=0}^{n-1} f^{-i}\alpha\right)$ for $f = \frac{2}{3}T$ or T is given at the bottom.

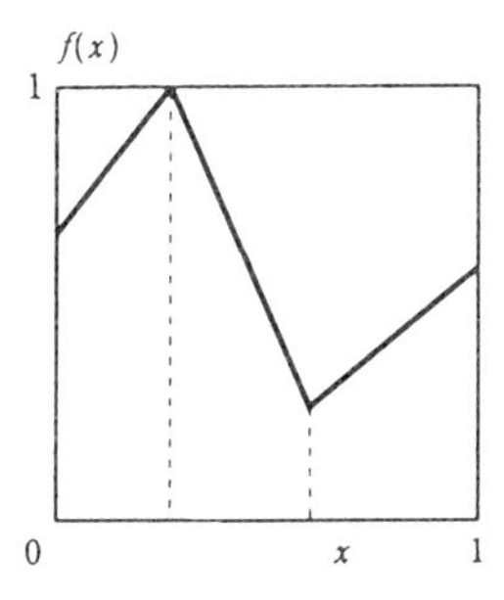

Figure 2.11. The lap number is 3 for this map ($\text{lap}(f) = 3$).

When the map f is piecewise-monotonic over the interval I, the topological entropy $h(f)$ may be determined by the *lap number* of the iterated map f^n:

$$h(f) = \lim_{n\to\infty} \frac{1}{n} \log \text{lap}(f^n). \tag{2.8}$$

Here the lap number $\text{lap}(f)$ is the least number of intervals in I over which f behaves monotonically (see figure 2.11). The lap number of f^n grows with n in general. If this growth obeys a power law, $\text{lap}(f^n) \sim cn^\alpha$ for example[9], one obtains

$$h(f) = \lim_{n\to\infty} \frac{1}{n} \log(cn^\alpha) = \lim_{n\to\infty} \frac{\alpha}{n} \log n = 0$$

while if it grows exponentially, $\text{lap}(f^n) \sim c\alpha^n$ $(\alpha > 1)$, one has

$$h(f) = \lim_{n\to\infty} \frac{1}{n} \log c\alpha^n = \log \alpha.$$

Thus, the topological entropy of a map f is determined by the way $\text{lap}(f^n)$ increases.

Let us explain how the topological entropy is related to the lap number with an example. Figure 2.12 is the graph of the map f^n for

$$f(x) = \begin{cases} x + \dfrac{1}{2} & \left(0 \le x \le \dfrac{1}{2}\right) \\[2ex] 2(1-x) & \left(\dfrac{1}{2} \le x \le 1\right) \end{cases}$$

which has a period 3 orbit $\{0, \frac{1}{2}, 1, 0, \frac{1}{2}, 1, \ldots\}$. The series of the lap numbers is $2, 3, 5, 8, 13, \ldots$, which is nothing but the Fibonacci series (a series defined by the recursion relation $a_n + a_{n+1} = a_{n+2}$ with $a_0 = 1$ and $a_1 = 2$ in the

[9] One writes $a_n \sim b_n$ when $\lim_{n\to\infty} a_n/b_n = 1$.

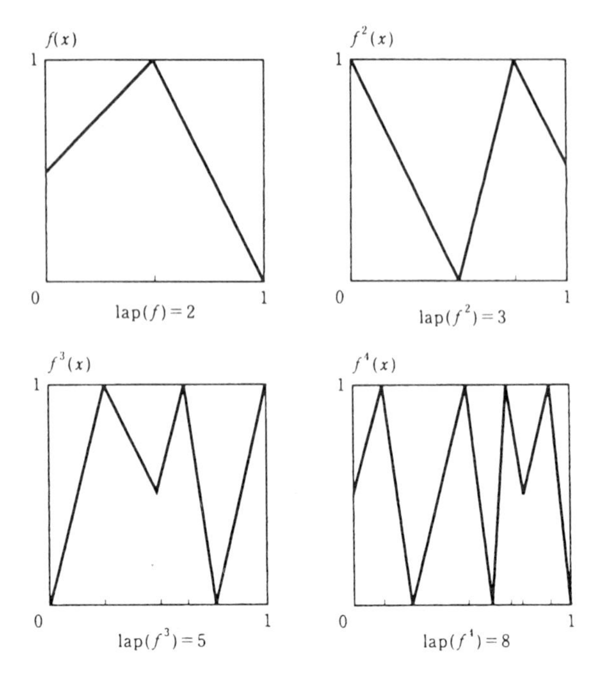

Figure 2.12. The lap number increases as a map is iterated.

present case). To make things clearer, let us cover $I = [0, 1]$ with a closed cover $\alpha = \{[0, \frac{1}{2}], [\frac{1}{2}, 1]\}$. Then, from

$$f^{-1}\left[0, \frac{1}{2}\right] = \left[\frac{3}{4}, 1\right] \qquad f^{-1}\left[\frac{1}{2}, 1\right] = \left[0, \frac{1}{2}\right] \cup \left[\frac{1}{2}, \frac{3}{4}\right]$$

one finds

$$f^{-1}\alpha = \left\{\left[0, \frac{1}{2}\right], \left[\frac{1}{2}, \frac{3}{4}\right], \left[\frac{3}{4}, 1\right]\right\} \qquad \alpha \vee f^{-1}\alpha = f^{-1}\alpha.$$

Now note that

$$N(\alpha) = 2 \quad N(f^{-1}\alpha) = 3.$$

Next, from

$$f^{-2}\alpha = \left\{\left[0, \frac{1}{4}\right], \left[\frac{1}{4}, \frac{1}{2}\right], \left[\frac{1}{2}, \frac{5}{8}\right], \left[\frac{5}{8}, \frac{3}{4}\right], \left[\frac{3}{4}, 1\right]\right\}$$

one obtains

$$\alpha \vee f^{-1}\alpha \vee f^{-2}\alpha = f^{-2}\alpha \qquad N\left(\bigvee_{i=0}^{2} f^{-i}\alpha\right) = N(f^{-2}\alpha) = 5 \ (= 2 + 3).$$

Furthermore one finds

$$N\left(\bigvee_{i=0}^{3} f^{-i}\alpha\right) = N(f^{-3}\alpha) = 8\ (= 3+5)$$

and eventually

$$N\left(\bigvee_{i=0}^{n-1} f^{-i}\alpha\right) = N(f^{-(n-1)}\alpha)$$

gives the Fibonacci series identical with $\mathrm{lap}(f^n)$.

Problem 6. Show for this example that $N(f^{-3}\alpha) = 8$.

It is often difficult to obtain the topological entropy $h(f)$ from its definition for a general map f and a number of methods to numerically compute $h(f)$ have been developed. In the following, we consider examples where $\mathrm{lap}(f^n)$ is obtained and $h(f)$ may be evaluated from this. The topological entropy of maps with superstable periodic points (see chapter 3) is calculated in appendix 3C.

(1) $L_R(x) = Rx(1-x)$ $(0 < R < 2)$. Since $L'_R(x) = 0$ only at $x = \frac{1}{2}$ and obviously $L^n_R(x) < \frac{1}{2}$, one finds from (2.2) that

$$(L^n_R(x))' = L'_R(L^{n-1}_R(x))L'_R(L^{n-2}_R(x))\ldots L'_R(L_R(x))L'_R(x) = 0$$

is satisfied if and only if $x = \frac{1}{2}$. Therefore the graph of $L^n_R(x)$ is essentially the same as $L_R(x)$ and hence $\mathrm{lap}(L^n_R) = 2$. Accordingly one easily finds $h(L_R) = 0$ and similarly $h(L_2) = 0$.

(2) $L_4 = 4x(1-x)$. The graph (figure 2.5(b)) clearly shows that $\mathrm{lap}(L^n_4) = 2^n$ and it follows from this that $h(L_4) = \log 2$. The tent map behaves in the same way.

(3) $L_R(x) = Rx(1-x)$ $(2 < R < 4)$. This case is rather complicated. First it should be noted from

$$(L^{n+1}_R(x))' = (RL^n_R(x)\{1 - L^n_R(x)\})' = R\{1 - 2L^n_R(x)\}(L^n_R(x))'$$

that the lap number satisfies[10]

$$\mathrm{lap}(L^{n+1}_R) = \mathrm{lap}(L^n_R) + \sharp\left\{x \,\middle|\, L^n_R(x) = \frac{1}{2}\right\}.$$

(i) $2 < R < 1+\sqrt{5} = 3.236\ldots$. This condition is equivalent to $L^2_R\left(\frac{1}{2}\right) = L_R(R/4) > \frac{1}{2}$ and then the number of solutions to $L^n_R(x) = \frac{1}{2}$ is found to be 2 for any n. It follows from this observation that $\mathrm{lap}(L^n_R)$ is an arithmetic series with the common difference $= 2$ and becomes $2n$ (see figure 2.13(a)). Then it follows that $h(L_R) = 0$.

[10] The symbol $\sharp\{A\}$ denotes the number of elements in a set A.

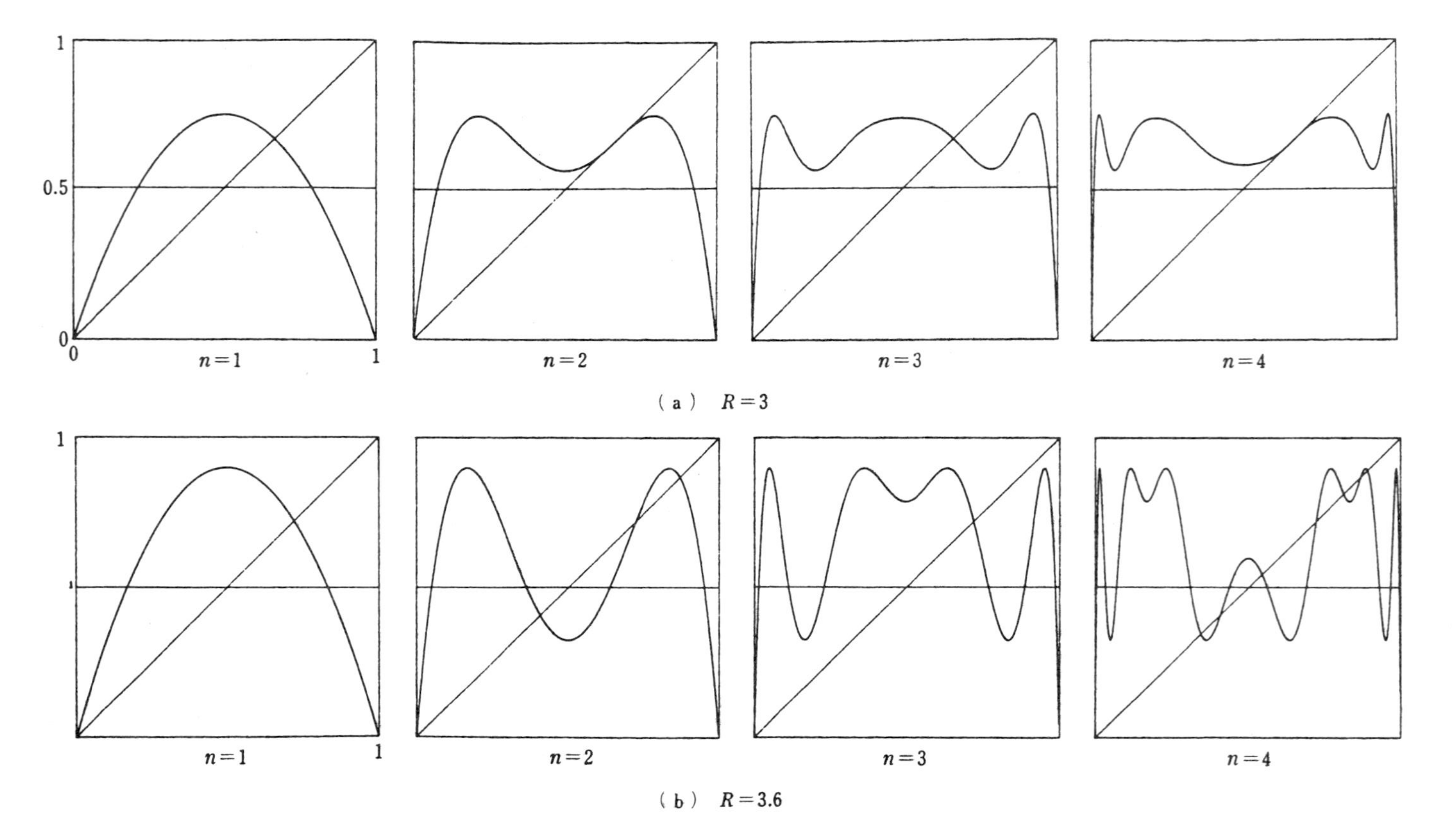

Figure 2.13. The iterated maps $L_R^n(x)$ of the map $L_R(x)$. The lap number increases as (a) $2n$ or (b) faster than $2n$ (grows exponentially for large n).

(ii) $1+\sqrt{5} < R$. $\sharp\{x \mid L_R^n(x) = \frac{1}{2}\}$ is greater than 2 and hence it is expected that the order of $\text{lap}(L_R^n)$ as a polynomial in n increases with R. (At this stage, the topological entropy $h(L_R)$ vanishes.) As R increases further, $\text{lap}(L_R)$ changes to an exponential function in n and $h(L_R) > 0$ (figure 2.13(b)). In fact, $\text{lap}(L_R^n) + 2$ turns out to be a Fibonacci series at $R = 3.8318\ldots$, which is a solution to $L_R^3(\frac{1}{2}) = \frac{1}{2}$, and one finds

$$\text{lap}(L_R^n) = A\left(\frac{1+\sqrt{5}}{2}\right)^n + B\left(\frac{1-\sqrt{5}}{2}\right)^n - 2 \qquad (A, B > 0)$$

from which one obtains

$$h(L_R) = \log\frac{\sqrt{5}+1}{2} = \log 1.618\ldots.$$

Problem 7. Show that $h(f) = \log\alpha$ when $\text{lap}(f^n) = A\alpha^n + B\beta^n + C$ ($A > 0$; $\alpha > 1$; $\alpha > |\beta| > 0$).

Let us comment on the relation between the topological entropy $h(f)$ of a map f and chaos. It is known that f is a Li–Yorke chaos provided that $h(f) > 0$. The converse is not true, however, and the existence of a Li–Yorke chaos f with vanishing $h(f)$ is known. In fact, the necessary and sufficient condition for $h(f) > 0$ is (see the end of section 2.3):

'f has a period $2^n(2m+1)$ orbit with $n \geq 0$ and $m \geq 1$'.

The necessary and sufficient condition for $h(f) > 0$ may take various different forms, one of which is concerned with the magnifying property of a map f:

'there exist disjoint closed intervals I and J and a natural number n such that $f^n(I) \cap f^n(J) \supset I$ and $f^n(I) \cap f^n(J) \supset J$'.

In the case of the tent map, for example, the width of an interval is doubled each time the map is applied and the condition above is clearly satisfied.

2.6 Denseness of orbits

2.6.1 Observable chaos and Lyapunov number

The existence of an uncountable scrambled set or the positivity of the topological entropy $h(f) > 0$ does not necessarily characterize the phenomenon called chaos as was mentioned in section 2.4. Let us consider the logistic map $L_R(x) = Rx(1-x)$ in the parameter range $3.5 < R < 4$. Figure 2.14(a)[11]

[11] Here the stationary orbits are defined as (1) the periodic orbits and (2) the orbits under an invariant measure. What are indicated under the name of chaotic orbits are those orbits that are based on the invariant measure which is most easily observable on a computer.

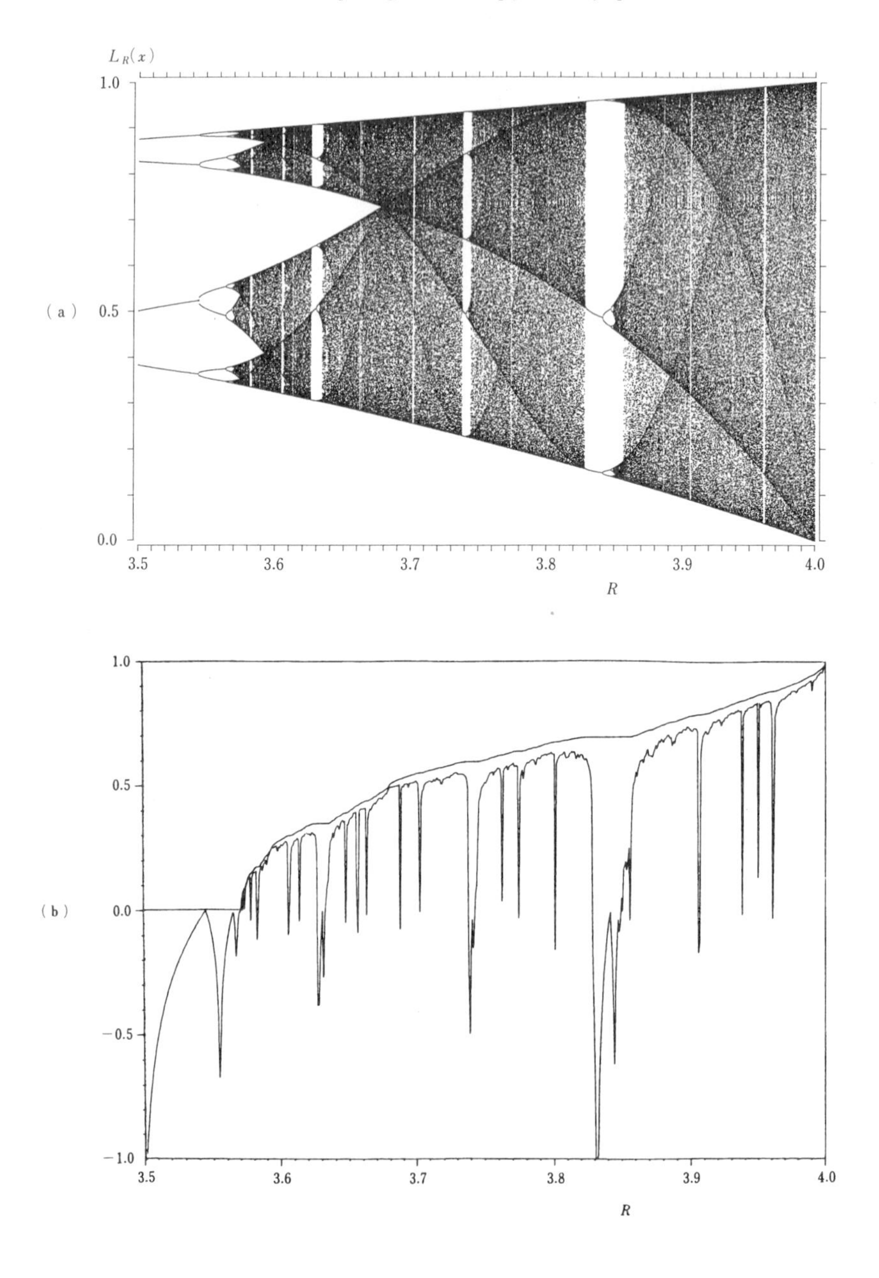

Figure 2.14. (a) The stationary orbits of $L_R(x) = Rx(1-x)$ as a function of R. (b) The topological entropy of $L_R(x)$ (the upper curve) and the Lyapunov number (the lower spiky curve). They are plotted with a common scale and both of their values at $R = 4$ are $\log 2$.

shows the orbits of L_R (the ordinate) as a function of R (the abscissa). There are stable period 3, 6, 12, ... orbits in the vicinity of $R = 3.84$, for example, and the Li–Yorke chaos, though it exists, is not observable. Such regions in R are called the *windows*. It looks as if the orbits of $L_R(x)$ for other $R(> R_\infty)$[12] are dense[13] in some intervals in x and as if the orbits are chaotic there. Figure 2.14(b) shows the topological entropy of L_R (the smooth curve) and the Lyapunov number[14] (the spiky curve), both computed numerically, for the same range of the parameter R. (The topological entropy has been obtained by the kneading sequence method outlined in appendix 3C.)

By comparing these two figures, one finds that (1) $h(f)$ monotonically increases after $R = R_\infty = 3.57\ldots$ (the bifurcation changes from 2^n-type period to $2^n(2m+1)$-type period at this value of R), where the chaotic behaviour takes place for the first time, while it is flat at the windows and (2) the Lyapunov number is negative at the windows. From these observations, one concludes that the Lyapunov number is more suited to identify observable chaos than the Li–Yorke chaos or the topological entropy. That is to say, the condition Lyapunov number > 0 implies observable chaos.

In reality, the Lyapunov number is related to the Kolmogorov entropy, which is different from the topological entropy. This aspect will not be mentioned in the present book.

It should be noted that the topological entropy of the logistic map montonically increases as a function of R (J Milnor and W Thurston [9]).

2.6.2 Denseness of orbits

The orbit of an observable chaos, when it appears, fills a certain region densely. This is related to the ergodic or the mixing property of the transformation, which will be explained in the following taking the binary transformation $B(x)$ as an example.

When a number $x \in [0, 1]$ is expressed in a binary fraction, it is a recurring fraction if x is rational while it is an infinite (nonrecurring) fraction if x is irrational as mentioned in 2.2.1. Since the binary transformation $B(x)$ shifts the binary fraction by one digit, the orbit of the transformation B with the initial point x is of period k if x is a recurring fraction whose repeating unit has the length k, while it is aperiodic if x is irrational. Accordingly an orbit $\{x, B(x), B^2(x), \ldots\}$ of the binary transformation with almost every initial point[15] $x \in [0, 1]$ is dense in this interval (that is, there are points belonging to this orbit in any interval

[12] It will be mentioned later that R_∞ is the supremum of R that generates the period 2^n, see section 3.1.

[13] See 2.6.2 for the meaning of 'dense'.

[14] See section 2.8.

[15] That is, all the numbers in [0, 1] but numbers belonging to a set with the measure zero. This is an expression often used in the Lebesgue integrals and the measure theory, see appendix 2D.

$(\alpha, \beta) \subset [0, 1]$, however small it may be). Moreover, they *distribute uniformly* in $[0, 1]$. In other words, if $N(x, (a, b), n)$ is the number of points, among the first n points $\{x, B(x), B^2(x), \ldots\}$ of this orbit, that belong to the interval (a, b), the uniformity of the distribution means that

$$\lim_{n\to\infty} \frac{1}{n} N(x, (a, b), n) = b - a$$

is true for any interval $(a, b) \subset [0, 1]$.

This will be explained with examples in the following. Let us divide the interval $[0, 1]$ into four pieces as shown in figure 2.15. The number in each subinterval is then characterized by two fractional digits; a number in $I_1 = (0, \frac{1}{4})$ takes the form $(0.00\ldots)_2$, a number in $I_2 = (\frac{1}{4}, \frac{1}{2})$, $(0.01\ldots)_2$, a number in $I_3 = [\frac{1}{2}, \frac{3}{4}]$, $(0.10\ldots)_2$ and finally a number in $I_4 = (\frac{3}{4}, 1)$, $(0.11\ldots)_2$.

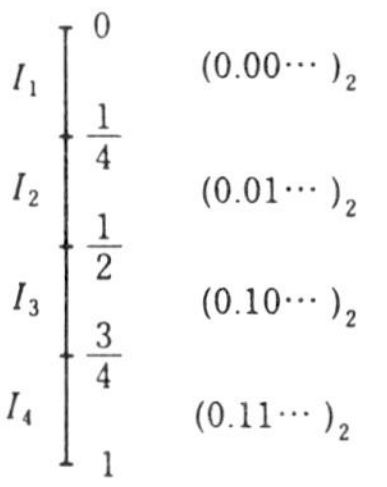

0
I_1 $(0.00\cdots)_2$
$\frac{1}{4}$
I_2 $(0.01\cdots)_2$
$\frac{1}{2}$
I_3 $(0.10\cdots)_2$
$\frac{3}{4}$
I_4 $(0.11\cdots)_2$
1

Figure 2.15. The interval [0, 1] divided into four and binary numbers.

Let us take a binary normal number (see appendix 2E) $x = 0.x_1x_2\ldots x_n\ldots$ and apply the binary transformation $B(x)$ repeatedly. Then the number is shifted as $B(x) = 0.x_2x_3\ldots x_{n+1}\ldots$, $B^2(x) = 0.x_3x_4\ldots x_{n+2}\ldots$. These numbers are found in the four subintervals with equal frequency. Similarly, if the interval $[0, 1]$ is divided into 2^n subintervals with equal length, the shifted series $\{x, B(x), B^2(x), \ldots, B^k(x), \ldots\}$ distribute over these subintervals with equal frequency. This means that an orbit of $B(x)$ with the initial point x distributes densely and uniformly in the interval $[0, 1]$. This is almost equivalent to the statement that x is a binary normal number (see problem 11).

Problem 8. The number $\frac{1}{\sqrt{2}}$ in the binary fraction form is

$$(0.101\,101\,010\,000\,010\,011\,110\,011\,001\,100\ldots)_2.$$

Find the number of points of the orbit $x, B(x), B^2(x), \ldots, B^{28}(x)$, contained in each of the four intervals I_1, I_2, I_3 and I_4 above.

Problem 9. Let $x = 0.x_1x_2\ldots x_n\ldots$ be a binary normal number. Show that the orbit of the binary transformation B with the initial point x is distributed uniformly over the interval $[0, 1]$.

Let us consider the tent map $T(x) = 1 - |2x - 1|$ next. This map reduces to $T(x) = 2x$ for $0 \le x < \frac{1}{2}$, being the same map as $B(x)$. For $\frac{1}{2} \le x < 1$ the tent map is $T(x) = 2 - 2x = 1 - B(x)$ since $B(x) = 2x - 1$ in this case. By noting that $1 = (0.111\ldots)_2$ and that $B(x)$ shifts a binary number by one digit, one finds that $T(x)$ shifts $x = 0.x_1x_2\ldots x_n\ldots$ toward the left by one digit to obtain $0.x_2x_3\ldots x_{n+1}\ldots$ followed by the replacement of $0 \leftrightarrow 1$. To be more explicit, this means

$$T(x) = 0.y_2y_3\ldots y_{n+1}\ldots \qquad (y_n = 1 - x_n, n = 2, 3, \ldots).$$

One concludes from this that the tent map $T(x)$ preserves the normal property of binary numbers. Therefore such properties of the binary transformation as the denseness of orbits and the uniformity of the distribution equally hold for the tent map.

Finally let us consider the logistic map $L(x) = 4x(1-x)$. If one substitutes

$$x_n = \sin^2\theta_n \tag{2.9}$$

into the relation $L(x_n) = x_{n+1} = 4x_n(1 - x_n)$, one finds

$$\theta_{n+1} = \begin{cases} 2\theta_n & \left(0 \le \theta_n \le \dfrac{\pi}{4}\right) \\ \pi - 2\theta_n & \left(\dfrac{\pi}{4} \le \theta_n \le \dfrac{\pi}{2}\right) \end{cases} \tag{2.10}$$

since $\sin^2\theta_{n+1} = 4\sin^2\theta_n(1 - \sin^2\theta_n) = \sin^2 2\theta_n$. This is just the tent map on the interval $\left[0, \frac{\pi}{2}\right]$. By introducing the variable $y_n = \frac{2}{\pi}\theta_n$, the logistic map becomes a tent map $y_{n+1} = T(y_n)$ defined by equation (1.3).

Problem 10. Show that the tent map $y_{n+1} = T(y_n)$ is obtained from the logistic map $x_{n+1} = L(x_n)$ by putting $y_n = \frac{2}{\pi}\theta_n = \frac{2}{\pi}\sin^{-1}\sqrt{x_n}$.

Thus the tent map is obtained from the logistic map and *vice versa* by a transformation of a variable. Such maps are called mutually *conjugate* and the properties of a map are found from those of the other map. For example, the denseness of orbits of the tent map mentioned above is also true for the logistic map.

Problem 11. Suppose orbits of the tent map are dense in the interval [0, 1]. Show that the orbit of the logistic map with the corresponding initial value is also dense in [0, 1]. (The property that the orbits are distributed uniformly over [0, 1] no longer holds.)

From $x_n = \sin^2\theta_n = \sin^2\frac{\pi}{2}y_n$ (problem 10), one has

$$\mathrm{d}x_n = 2\sin\theta_n\cos\theta_n\mathrm{d}\theta_n = 2\sqrt{x_n(1 - x_n)}\frac{\pi}{2}\mathrm{d}y_n$$

and hence

$$\mathrm{d}y_n = \frac{1}{\pi\sqrt{x_n(1-x_n)}}\mathrm{d}x_n. \tag{2.11}$$

Accordingly the points of an orbit of the logistic map are not distributed uniformly over [0, 1] but distributed with more points on the both ends, even though orbits of the tent map are distributed uniformly. This is clearly seen in figure 1.12. It should be also noted that the initial value of the binary transformation or the tent map giving a periodic orbit is rational while it becomes irrational for the logistic map under the transformation $x = \sin^2 \frac{\pi}{2} y$. This is in consistent with the situation where orbits are generated with these maps. That is, a simple rational initial value such as 0.2 or 0.35 generates a periodic orbit in the binary transformation or the tent map while it generates an aperiodic orbit in the logistic map.

2.7 Invariant measure

In the previous section, we considered the distribution of the points $\{x_n\}$ of an orbit of the binary transformation, the tent map and the logistic map starting with a certain initial value. It was shown there that the distribution of the points defines a distribution function $\rho(x)$ as n becomes large. It is convenient to normalize $\rho(x)$ over [0, 1] as $\int_0^1 \rho(x)\mathrm{d}x = 1$:

$$\rho(x) = \lim_{N\to\infty} \frac{1}{N} \sum_{i=1}^{N} \delta(x - x_i). \tag{2.12}$$

Here $\delta(x)$ is the Dirac delta function (appendix 2G).

For a period k orbit, this becomes

$$\rho(x) = \frac{1}{k} \sum_{i=1}^{k} \delta(x - x_i). \tag{2.13}$$

If one starts with a randomly irrational number (a normal number in appendix 2E) in an aperiodic orbit of $B(x)$ or $T(x)$, one obtains

$$\rho(x) = 1. \tag{2.14}$$

For an aperiodic orbit of $L(x)$ corresponding to an aperiodic orbit of $B(x)$, one obtains the distribution function

$$\rho(x) = \frac{1}{\pi} \frac{1}{\sqrt{x(1-x)}}. \tag{2.15}$$

The function $\rho(x)$ is called the density of the *invariant measure*. The average $\langle A(x)\rangle$ of a quantity $A(x)$ with respect to an orbit is written with the help of the invariant measure $\rho(x)$ as a weighted spatial average as

$$
\begin{aligned}
\langle A(x)\rangle &= \lim_{N\to\infty}\frac{1}{N}\sum_{i=1}^{N}A(x_i) = \lim_{N\to\infty}\frac{1}{N}\sum_{i=1}^{N}\int A(x)\delta(x-x_i)\mathrm{d}x \\
&= \int A(x)\lim_{N\to\infty}\frac{1}{N}\sum_{i=1}^{N}\delta(x-x_i)\mathrm{d}x \\
&= \int A(x)\rho(x)\mathrm{d}x
\end{aligned}
\tag{2.16}
$$

where the integration is carried out over the interval I.

The invariant measure $\rho(x)$ of the binary transformation $B(x)$ and the tent map $T(x)$ is classified into three cases according to the initial condition as

(1) periodic orbits (including eventually periodic cases)
(2) randomly irrational cases and
(3) nonrandomly irrational cases.

Among the three cases above, (2) exhausts almost all cases and the corresponding invariant measure is given by equation (2.14). For the map $L(x)$, which is conjugate to $T(x)$, the corresponding invariant measure is given by equation (2.15).

A final remark on the relation that the invariant measure $\rho(x)$ satisfies is in order. Let $f(x)$ be a map. Then it follows from the identity

$$
\int \delta(y-f(x))\delta(x-x_i)\mathrm{d}x = \delta(y-f(x_i)) = \delta(y-x_{i+1})
$$

that

$$
\begin{aligned}
\rho(y) &= \lim_{N\to\infty}\frac{1}{N}\sum_{i=1}^{N}\delta(y-x_i) = \lim_{N\to\infty}\frac{1}{N}\sum_{i=1}^{N}\delta(y-x_{i+1}) \\
&= \lim_{N\to\infty}\frac{1}{N}\sum_{i=1}^{N}\int \delta(y-f(x))\delta(x-x_i)\mathrm{d}x \\
&= \int \delta(y-f(x))\lim_{N\to\infty}\frac{1}{N}\sum_{i=1}^{N}\delta(x-x_i)\mathrm{d}x.
\end{aligned}
$$

Thus one finally obtains

$$
\rho(y) = \int \delta(y-f(x))\rho(x)\mathrm{d}x. \tag{2.17}
$$

Here the integration domain is restriced to the domain of the map. Equation (2.17) is called the *Frobenius–Perron relation.*

Problem 12. Show that (1) $f(x) = T(x) = 1 - |2x - 1|$ with $\rho(x) = 1$ and (2) $f(x) = L(x) = 4x(1 - x)$ with

$$\rho(x) = \frac{1}{\pi} \frac{1}{\sqrt{x(1-x)}}$$

both satisfy the Frobenius–Perron relation.

2.8 Lyapunov number

We have studied maps $L(x), T(x)$ and $B(x)$ generating chaos so far. They are characterized by the property that the inverse map is double valued, which makes the average gradient of the map larger than unity. This large gradient is the origin of the fundamental characteristics of chaos, namely *stretching* and *folding*, mentioned in chapter 1. Accordingly one may characterize chaos by the condition that the average gradient of the map is greater than unity or that the distance of two nearby points increases exponentially with time. The long time average of the exponent with respect to an orbit is denoted by λ and called the *Lyapunov number*:

$$\lambda = \lim_{N \to \infty} \frac{1}{N} \sum_{i=0}^{N-1} \log |f'(x_i)|. \tag{2.18}$$

In other words, the distance of two nearby orbits mentioned above increases exponentially if $\lambda > 0$. A system whose orbit satisfies this property is called an *unstable system*.

Since $|f'(x)| = 2$ for $T(x)$ and $B(x)$, one has $\lambda = \log 2$ for these maps. If the distribution of an irregular orbit $\{x_i\}$ is nonuniform over the interval $[0, 1]$, the Lyapunov number is computed following equation (2.16) with the invariant measure $\rho(x)$ as

$$\lambda = \int \log |f'(x)| \rho(x) \mathrm{d}x. \tag{2.19}$$

Let us compute the Lyapunov number λ for the logistic map explicitly. Substituting $f'(x) = 4 - 8x$ and $\rho(x) = 1/(\pi\sqrt{x(1-x)})$ into equation (2.19), one has

$$\lambda = \int_0^1 \frac{\log |4 - 8x|}{\pi\sqrt{x(1-x)}} \mathrm{d}x = 2 \int_0^{1/2} \frac{\log\{4(1-2x)\}}{\pi\sqrt{x(1-x)}} \mathrm{d}x.$$

After the change of the variable $x = \sin^2 \theta$ and $\mathrm{d}x = 2 \sin\theta \cos\theta \mathrm{d}\theta$, one obtains

$$\begin{aligned}
\lambda &= 2 \int_0^{\pi/4} \frac{\log\{4(1 - 2\sin^2\theta)\}}{\pi \sin\theta \cos\theta} 2 \sin\theta \cos\theta \mathrm{d}\theta \\
&= \frac{4}{\pi} \int_0^{\pi/4} \log(4 \cos 2\theta) \mathrm{d}\theta \\
&= \log 4 + \frac{4}{\pi} \int_0^{\pi/4} \log(\cos 2\theta) \mathrm{d}\theta
\end{aligned}$$

$$= \log 4 + \frac{2}{\pi} \int_0^{\pi/2} \log(\cos\theta')\mathrm{d}\theta' \qquad \text{(where } \theta' = 2\theta\text{)}.$$

The second term in the last line is $-\log 2$ and one finally finds $\lambda = \log 2$.

Problem 13. Show that

$$I = \frac{2}{\pi} \int_0^{\pi/2} \log(\cos\theta)\mathrm{d}\theta = -\log 2.$$

(This integral is called the Euler integral.)

The positivity of the Lyapunov number is most often employed as a criterion for chaos in the analysis of an irregular wave generated by a computer, experiment or observation. In fact, this condition is an excellent tool in judging an oscillatory wave being irregular or an orbit being complex.

It should be also mentioned that the Lyapunov number is applicable not only to one-dimensional maps studied so far but to more general maps and systems of differential equations. It will be mentioned in detail in chapter 4 that there are n Lyapunov numbers ($\lambda_1, \lambda_2, \ldots, \lambda_n$) for a system with n variables. Among these n Lyapunov numbers, the largest is called the *maximum Lyapunov number* and a system is said to be chaotic, in many cases, if the maximum Lyapunov number is positive. However, it should be noted that, when the number of the dimension is extremely large, the positivity of the maximum Lyapunov number does not necessarily characterize complex behaviour that is seen in a small-dimensional system. We note *en passant* that the computation of the topological entropy is more complicated than that of the Lyapunov number. Therefore there are not many evaluations of the topological entropy except for one-dimensional maps.

2.9 Summary

Let us summarize the Li–Yorke chaos, the topological entropy $h(f)$ and the Lyapunov number λ from the viewpoint of the condition for the existence of chaos in a one-dimensional map.

(1) The necessary and sufficient condition for $h(f) > 0$ is that the map f has a period $2^n(2m+1)$, $(n \geq 0, m \geq 1)$.
(2) There exists a scrambled set in the sense of Li–Yorke if $h(f) > 0$. In other words, the map f is Li–Yorke chaos.
(3) The condition $\lambda > 0$ is the criterion for the existence of observable chaos.
(4) The condition $h(f) > 0$ may be used for the criterion for the existence of chaos, both observable and unobservable.
(5) The condition $h(f) = 0$ implies, in most cases, a nonchaotic orbit, which may be a periodic orbit, while it leads to Li–Yorke chaos in other cases.

The condition $h(f) > 0$ with (i) $\lambda < 0$ implies the existence of potential chaos while with (ii) $\lambda > 0$ it means observable chaos.

Chapter 3

Routes to chaos

A system showing chaotic behaviour undergoes transitions between nonchaotic and chaotic states in general. There are several ways in which a system undergoes a transition to chaos; three typical ones are:

(a) Through consecutive *pitchfork bifurcations* to chaos. This is commonly called the *Feigenbaum route.*

(b) Through inverse *tangent bifurcations* or *intermittency chaos* to chaos. This route to chaos is called the *Pomeau–Manneville route.*

(c) Through repeated *Hopf bifurcations* to chaos. This is a route to chaos stressed by Ruelle and Takens.

Here (a) and (b) will be explained since they appear often in one-dimensional maps.

3.1 Pitchfork bifurcation and Feigenbaum route

Let us introduce a parameter in the logistic map $L(x)$ and the tent map $T(x)$ as we did in chapter 2:

$$T_A(x) \equiv AT(x) = A(1 - |2x - 1|) \quad (0 < A \leq 1) \tag{3.1}$$

$$L_R(x) \equiv \frac{R}{4}L(x) = Rx(1 - x) \quad (0 < R \leq 4). \tag{3.2}$$

These parametrized maps $L_R(x)$ and $T_A(x)$ will be also called the logistic map and the tent map as before. The domain of x is also taken to be $[0, 1]$.

Let us consider the tent map first. Figure 3.1 shows that all the orbits starting within the domain approach 0 asymptotically when $A < 1/2$. Therefore chaos is not produced. In case $A > 1/2$, in contrast, there appears an unstable fixed point at $x = 2A/(2A + 1)$ and, at the same time, the point $x = 0$ becomes unstable as well. The slope of the map is greater than 1 and the Lyapunov

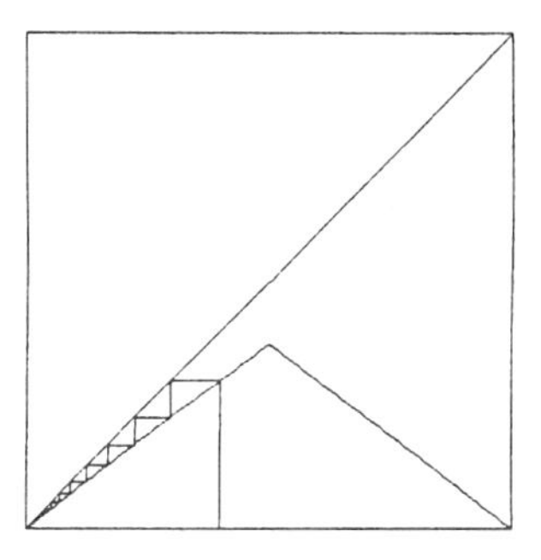

Figure 3.1. An orbit of the tent map $T_A(x)$ with $A < \frac{1}{2}$. It approaches 0 asymptotically. ($A = 0.375$.)

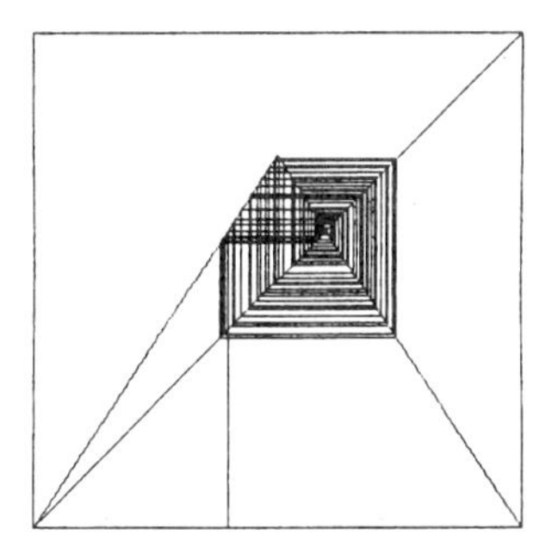

Figure 3.2. A chaotic orbit of the tent map $T_A(x)$ with $A > \frac{1}{2}$. The orbit is distributed in the interval $[2A(1 - A), A]$. ($A = 0.75$.)

number is a positive number $\log 2A$ and hence these facts both lead to chaos. In fact, an aperiodic orbit, that is characteristic of chaos, is observed in numerical computation as shown in figure 3.2.

It should be noted that the aperiodic orbits in this case are not distributed everywhere in the interval $[0, 1]$, but localized within the interval $[2A(1-A), A]$. Moreover, the orbits are further localized within parts of the above interval for $\frac{1}{2} < A < \frac{\sqrt{2}}{2}$, while the orbit spreads throughout the interval for $\frac{\sqrt{2}}{2} < A \leq 1$ as shown in figure 3.3 [10].

The topological entropy of the tent map is $\log 2A$ when $A \geq \frac{1}{2}$, which agrees with the Lyapunov number (Ito, Tanaka and Nakada [10]).

Problem 1. Consider orbits of the tent map $T_A(x)$ with $A > \frac{1}{2}$. Show that any orbit whose initial point lies in the interval $(0, 1)$ takes its value in $[2A(1-A), A]$.

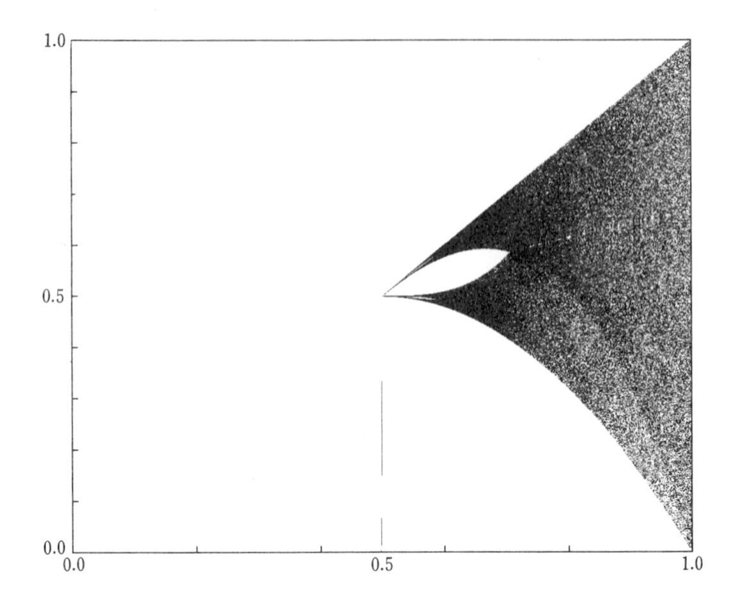

Figure 3.3. The stationary orbits of the tent map $T_A(x)$.

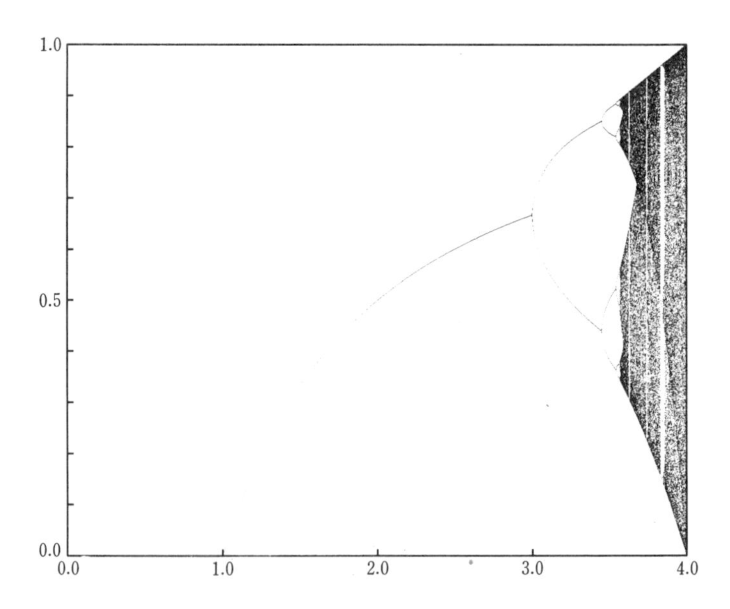

Figure 3.4. The stationary orbits of the logistic map $L_R(x)$.

In summary, the stationary orbit (i.e. an orbit after a large number of iterations) of the tent map makes a sudden transition at $A = \frac{1}{2}$ from a stable periodic orbit $x = 0$ to a chaotic orbit as A is increased. If A is further increased, the chaotic region spreads throughout the interval [0, 1] as shown in figure 3.3.

In the logistic map (3.2), in contrast, there appear various stationary orbits as R is increased as shown in figure 3.4. There appears a sequence of pitchfork bifurcations shown in figure 3.7 in this map if R takes a value in the range $3 \sim 3.57 (\simeq R_\infty)$. Here *bifurcation* means that a stable solution becomes unstable as the parameter changes and there appears, at the same time, a new stable solution.

It turns out from numerical computations that these consecutive pitchfork bifurcations appear infinitely many times and the bifurcation point approaches R_∞ as a geometric progression. As the parameter R is increased, the period of the stable periodic solution doubles as $1, 2, 4, 8, \ldots$, which is called the *period doubling phonomenon*. Finally there appears the chaotic region with $R > R_\infty$.

The orbit distributes in a certain range of x when $R > R_\infty$, showing chaotic behaviour. For certain ranges of $R > R_\infty$, however, there appear *windows* where chaos does not exist. These various phenomena in the logistic map, which are not seen in the tent map, are due to the absence of a sharp vertex and to the upward convexity of the whole curve. They will be treated in detail in sections 3.2 and 3.3.

The pitchfork bifurcations and the period doubling phonomena of the logistic map will be explained in the following.

Suppose R is gradually increased.

(1) $R < 1$. The intersection point (i.e. fixed point) of $L_R(x)$ and the diagonal line ($y = x$) is $x = 0$ only and this point is stable; an orbit starting from any point in [0, 1] approaches $x = 0$. This is similar to the case $A < \frac{1}{2}$ in the tent map (see figure 3.5).

(2) $1 < R < 3$. The condition $L_R(x) = x$ yields two fixed points $x_0 = 0$ and

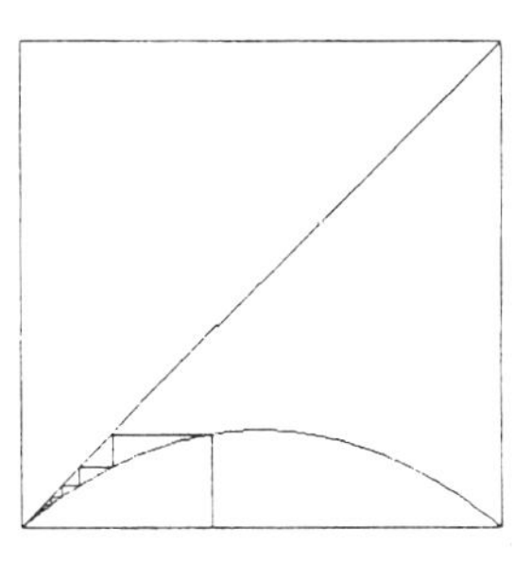

Figure 3.5. An orbit of the logistic map $L_R(x)$ with $R < 1$. It asymptotically approaches $x = 0$. Here $R = 0.8$.

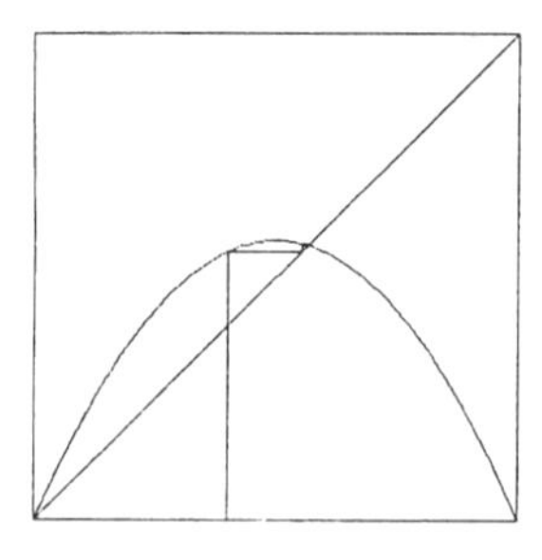

Figure 3.6. An orbit of the logistic map $L_R(x)$ with $1 < R < 3$. It asymptotically approaches $x = 1 - \frac{1}{R}$. Here $R = 2.3$.

$x_1 = 1 - \frac{1}{R}$. The point x_0 is unstable since $|f'(x_0)| = R > 1$ while x_1 is stable. In fact, figure 3.6 shows that the orbit approaches x_1.

(3) $3\ (=R_1) < R < 1+\sqrt{6}\ (=R_2)$. The inequality $|L_R'(x)| > 1$ implies that the fixed point x_1 is no longer stable. Then there appear two stable period 2 points x_{2-} and x_{2+} on both sides of x_1 as shown in figure 3.7. This is the pitchfork bifurcation and the bifurcation point is $R = 3$. The points x_{2-} and x_{2+} are obtained by solving the equation $L_R^2(x) = x$ giving the fixed point of the twice-iterated map $L_R^2(x)$. It follows from

$$L_R^2(x) - x = -x\{Rx - (R-1)\}\{R^2x^2 - R(R+1)x + (R+1)\} = 0$$

that there appear new solutions

$$x_{2\pm} = \frac{R + 1 \pm \sqrt{(R+1)(R-3)}}{2R} \tag{3.3}$$

in addition to x_0 and x_1. These new solutions are stable in the range of $R_1 < R < R_2$. This is because

$$\frac{\mathrm{d}}{\mathrm{d}x}L_R^2(x) = \frac{\mathrm{d}}{\mathrm{d}x}L_R(L_R(x)) = \frac{\mathrm{d}L_R(L_R(x))}{\mathrm{d}L_R(x)}\frac{\mathrm{d}L_R(x)}{\mathrm{d}x}$$

and one obtains

$$(L_R^2)'(x_{2+}) = L_R'(x_{2+})L_R'(x_{2-})$$

if one puts $x = x_{2+}$ and notices $L_R(x_{2+}) = x_{2-}$. Thus it follows from $L_R(x_{2-}) = x_{2+}$ that

$$\begin{aligned}(L_R^2)'(x_{2-}) &= L_R'(x_{2+})L_R'(x_{2-}) = R^2(1 - 2x_{2+})(1 - 2x_{2-}) \\ &= (L_R^2)'(x_{2+}) = -R^2 + 2R + 4\end{aligned}$$

whose absolute value is less than unity for $3 < R < 1 + \sqrt{6}$.

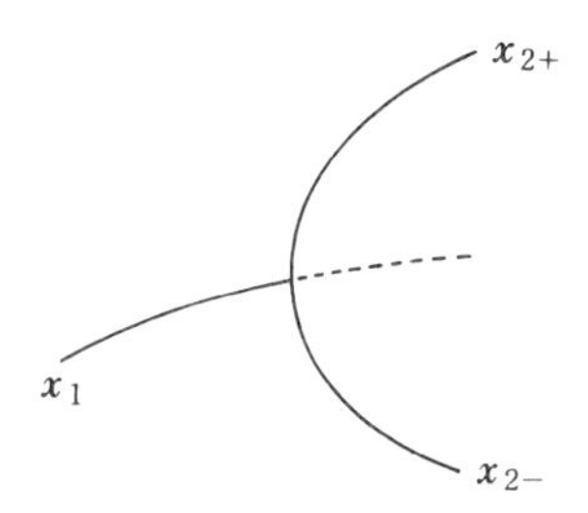

Figure 3.7. A pitchfork bifurcation. The solid lines are stable orbits while the broken line is an unstable orbit. Stable orbits are observed as stationary orbits.

Problem 2. Prove the above inequality.

(4) $R > R_2 = 1 + \sqrt{6}$. The points x_{2+} and x_{2-} become unstable and there appear stable period 4 solutions. Since

$$(L_R^2)'(x_{2+}) = (L_R^2)'(x_{2-}) = -1$$

at the bifurcation point $R = R_2 = 1 + \sqrt{6}$, the points $x_{2\mp}$ are destablized simultaneously and each of them bifurcates to two stable solutions. The way the solutions $x_{2\pm}$ of $L_R^2(x)$ bifurcate is the same as the way that the solution x_1 of $L_R(x)$ is destablized to produce $x_{2\pm}$ (see figure 3.8). The stable period 4 solutions for $R > 1+\sqrt{6}$ are found as the fixed points of the four-times-iterated map $L_R^4(x)$. These period 4 points cannot be obtained, however, by algebraic manipulations. This is because the equation to be solved is $L_R^4(x) - x = 0$, which is of degree $2^4 = 16$ in x. Even when the known solutions x_0, x_1 and $x_{2\pm}$ are factored out, the rest is still of degree 12, for which no formula for the solutions is known. Figure 3.9 shows the period 2, 4, 8 and 16 solutions of the logistic map.

Problem 3. Show that the polynomial $L_R^4(x) - x$ has factors $L_R^2(x) - x$ and $L_R(x) - x$.

As R is further increased, period 2^{n-1} points are destablized at $R = R_n$ and stable points with period 2^n appear through pitchfork bifurcations.

The distance between successive R_n becomes smaller as a geometrical progression for large n. Let $1/\delta$ be its common ratio. This problem has been analysed by M J Feigenbaum [11], who found that

$$\delta = \lim_{n\to\infty} \frac{R_n - R_{n-1}}{R_{n+1} - R_n} = 4.669\,201\,609\ldots. \tag{3.4}$$

The number δ is called the Feigenbaum constant. R_n approaches $R_\infty = 3.569\,9456\ldots$ according to numerical computations. By changing the abscissa

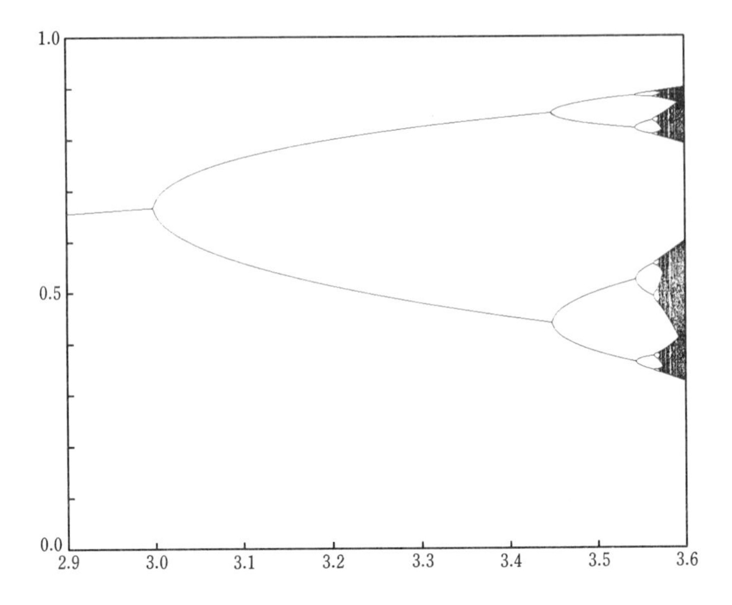

Figure 3.8. The range of R on which the logistic map has pitchfork bifurcations. The map is chaotic above $R_\infty = 3.569\,9456\ldots$.

of figure 3.8 to $-\log(R_\infty - R)$ to see the distribution of $\{R_n\}$, one obtains figure 3.10, which shows the distribution is almost equidistant even for small n. The number δ of equation (3.4) is obtained from numerical calculations and is an asymptotic value for large n. It is interesting, however, to estimate δ from R_n with small n. Let us employ R_0, R_1 and R_2 as the bifurcation points. In spite of the fact that R_0 is the bifurcation point of the stable solution $x = 0$ and hence it is questionable to say it is of the pitchfork type, it is considered essentially to be a pitchfork bifurcation since the instability appears because the map $L_R(x)$ is tangent to the diagonal line $x_{n+1} = x_n$. It is surprising that the number

$$\delta_1 = \frac{R_1 - R_0}{R_2 - R_1} = \frac{3 - 1}{(1 + \sqrt{6}) - 3} \simeq 4.45 \tag{3.5}$$

obtained from these bifurcation points is fairly close to the correct δ.

Suppose R_n converges to R_∞ with this ratio. Then one estimates R_∞ as

$$\begin{aligned} R_\infty - R_1 &= (R_\infty - R_2) + (R_2 - R_1) = \ldots \\ &= (R_2 - R_1) + (R_3 - R_2) + (R_4 - R_3) + \ldots \\ &= (R_2 - R_1)\left\{1 + \frac{1}{\delta_1} + \frac{1}{\delta_1^2} + \ldots\right\} = (R_2 - R_1)\frac{\delta_1}{\delta_1 - 1}. \end{aligned}$$

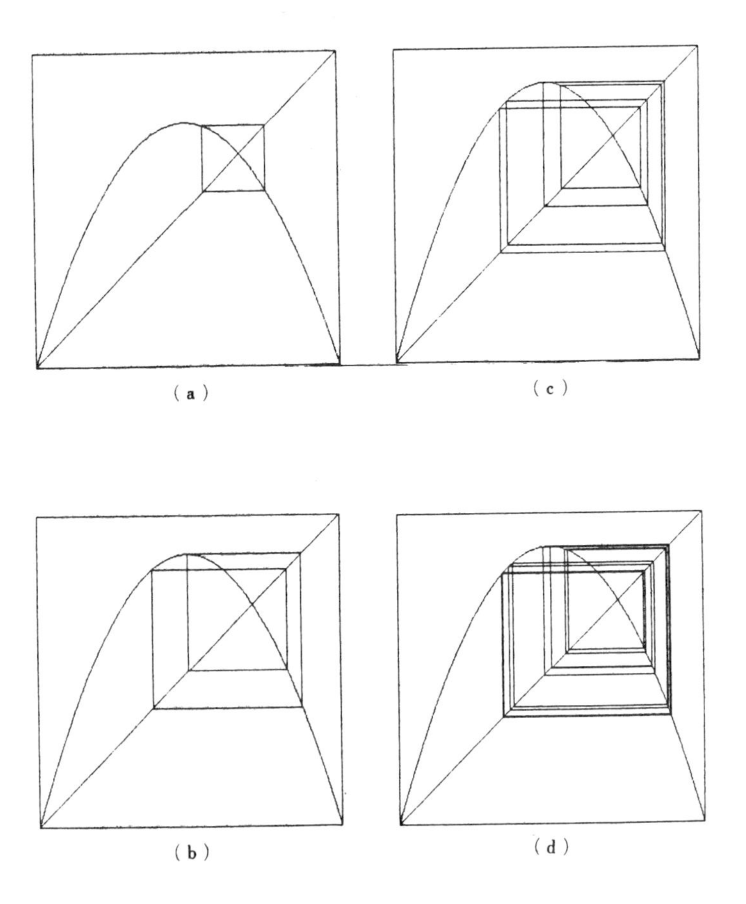

Figure 3.9. Periodic points of the logistic map. (a) The period 2 points ($R = 3.1$). (b) The period 4 points ($R = 3.5$). (c) The period 8 points ($R = 3.56$). (d) The period 16 points ($R = 3.5685$).

Thus the estimated R_∞ is

$$R_\infty = R_1 + (R_2 - R_1)\frac{\delta_1}{\delta_1 - 1} = 3.580 \qquad (3.6)$$

which is again very close to the actual value

$$R_\infty = 3.569\,9456\ldots$$

obtained from numerical computations. The lesson one learns from these observations is that the distribution of the sequence R_n of the logistic map is close to a geometrical progression specified by the Feigenbaum number δ even

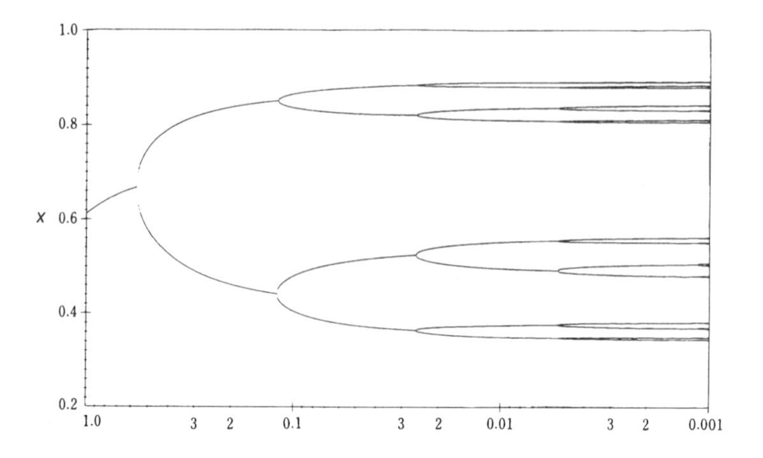

Figure 3.10. The pitchfork bifurcations of the logistic map are plotted with $-\log(R_\infty - R)$ as the abscissa.

for small n. More generally, consider a continuous map

$$x_{n+1} = Rf(x_n). \tag{3.7}$$

If $f(x_n)$ has a single peak and the behaviour of $f(x_n)$ near the peak is quadratic (namely $f''(x) \neq 0$) and, furthermore, it satisfies the Schwarz condition mentioned in the next section, the bifurcation points $\{R_n\}$ converge to R_∞ as a geometrical progression with the common ratio $1/\delta$, where the constant δ is given by equation (3.4).

Moreover, there exists a distinctive property of the Feigenbaum route. That is, as the map is iterated many times, the shape of the curve near the peak, with a similarly reduced scale, approaches a fixed curve $g(x)$ (see figure 3.11). This is expressed mathematically as

$$\lim_{n\to\infty} (-\alpha)^n f^{2n}_{R_{n+1}} \left[\frac{x'}{(-\alpha)^n}\right] = g(x') \tag{3.8}$$

where $x' = x - \frac{1}{2}$,

$$g(x') = 1 - 1.527\,63x'^2 + 0.104\,815x'^4 - 0.026\,7057x'^6 + \dots \tag{3.9}$$

and

$$\alpha = 2.502\,807\,876\dots.$$

An estimate of this number α from the bifurcation points with small n is

$$\frac{x_1 - \dfrac{1}{2}}{\dfrac{1}{2} - x_{2-}} = \left.\frac{\left(1 - \dfrac{1}{R}\right) - \dfrac{1}{2}}{\dfrac{1}{2} - \dfrac{R + 1 - \sqrt{(R+1)(R-3)}}{2R}}\right|_{R=R_\infty} = 2.557\dots \tag{3.10}$$

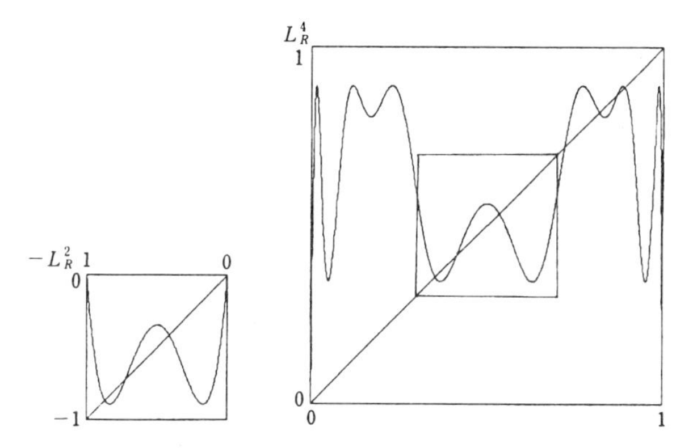

Figure 3.11. The graphs of $-L_R^2(x)$ with $R = 3$ and $L_R^4(x)$ with $R = 3.449\,49$. The graph of $-L_R^2(x)$ is reduced by 1/2.5. The central part of $L_R^4(x)$ is very similar to $-L_R^2(x)$.

which is fairly close to the true value of α in spite of the smallness of n.

Problem 4. Show that

$$g(x) = Tg(x) \equiv -\alpha g\left[g\left(-\frac{x}{\alpha}\right)\right] \tag{3.11}$$

is true for $x = 0$, provided that $g(x)$ is given by equation (3.9).

Problem 5. Find an approximate value of α by putting $g(x) = 1 + Ax^2$ in equation (3.11).

The Feigenbaum route to chaos based on the infinite sequence of pitchfork bifurcations is seen in a wide class of maps as mentioned before. Figure 3.12 shows an example, $x_{n+1} = A\sin(\pi x_n)$, of this class. Note that the way the orbit bifurcates is very similar to that of the logistic map.

3.2 Condition for pitchfork bifurcation

Let us consider what conditions a map $f(x)$ must satisfy for the existence of (1) the infinite sequence of pitchfork bifurcations and (2) the Feigenbaum ratio δ mentioned in section 3.1.

We consider (2) first. A higher order bifurcation is controlled by the shape of the peaks for higher order map $L_R^n(x)$ with $n \gg 1$. This is because $\tilde{x} = \frac{x'}{(-\alpha)^n}$ takes a value close to 0 when x' changes from $-\frac{1}{2}$ to $\frac{1}{2}$ and it is known that equation (3.4) follows if $f(x)$ has a single peak and behaves like a quadratic function ($f''(0) \neq 0$) near the peak as the logistic map.

Consider (1) next. Let us first consider the condition for the existence of a pitchfork bifurcation. Period 2 points are destablized at R_2 where stable period

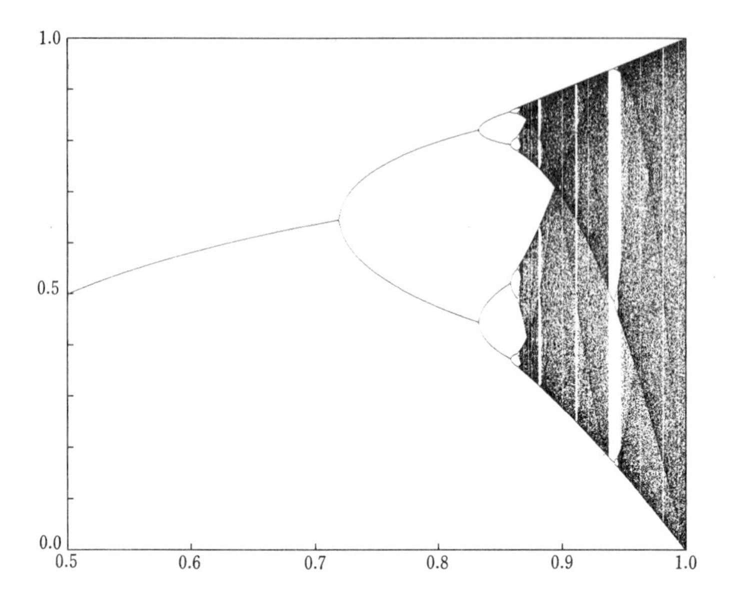

Figure 3.12. The bifurcation diagram of the map $x_{n+1} = A\sin(\pi x_n)$.

4 points appear and these period 4 ($= 2^2$) points are destablized at R_3. Generally speaking, stable 2^n points are destablized at $R = R_{n+1}$. Similarly to the case of $R = R_2$ mentioned in section 3.1, one has

$$(L^{2^n}_{R_{n+1}})'(x_k) = -1 \qquad (k = 1, 2, \cdots, 2^n) \tag{3.12}$$

where x_k is the kth fixed point of $L^{2^n}_{R_{n+1}}(x) = x$ to be destablized (see figure 3.13).

Problem 6. Show by mathematical induction that there are 2^{n+1} fixed points of $L^{2^n}_R(x) = x$ for $R_n < R < R_{n+1}$ and 2^n of them are stable fixed points while the other 2^n are unstable fixed points.

Put $L^{2^n}_{R_n}(x) = f_R(x)$ and $L^{2^{n+1}}_{R_n}(x) = f_R(f_R(x)) = g_R(x)$. It follows from $f'_R(x) = -1$ and

$$g'(x_k) = \left.\frac{\mathrm{d}f^2(x)}{\mathrm{d}f(x)}\frac{\mathrm{d}f(x)}{\mathrm{d}x}\right|_{x=x_k} = (f'(x_k))^2 = 1$$

$$g''(x_k) = \left.\left\{\frac{\mathrm{d}^2 f^2(x)}{\mathrm{d}(f(x))^2}\left(\frac{\mathrm{d}f(x)}{\mathrm{d}x}\right)^2 + \frac{\mathrm{d}f^2(x)}{\mathrm{d}f(x)}\frac{\mathrm{d}^2 f(x)}{\mathrm{d}x^2}\right\}\right|_{x=x_k} = 0 \tag{3.13}$$

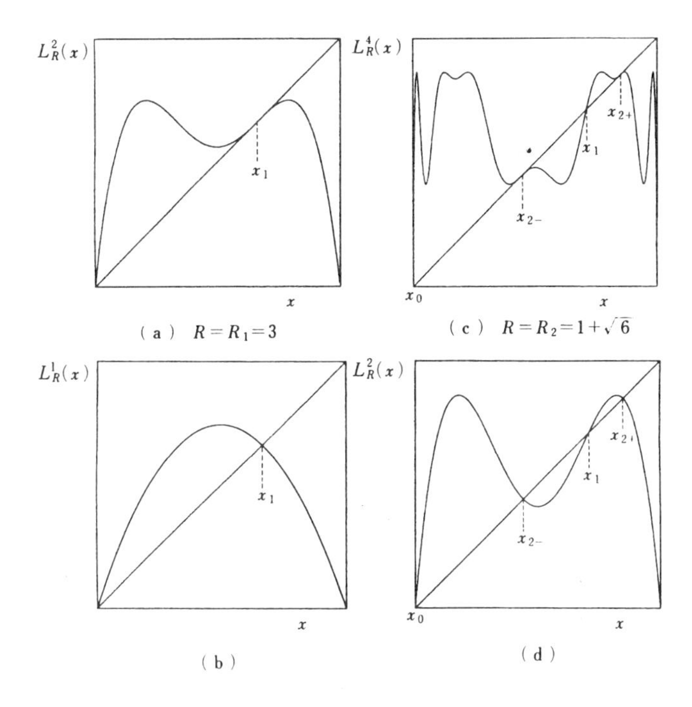

Figure 3.13. (a) $L_R^2(x)$ and (b) $L_R^1(x)$ at the destabilizing point $R = R_1$ of the period 1 point $x = x_1$. (c) $L_R^4(x)$ and (d) $L_R^2(x)$ at the destabilizing point $R = R_2$ of the period 2 points $x = x_{2\pm}$.

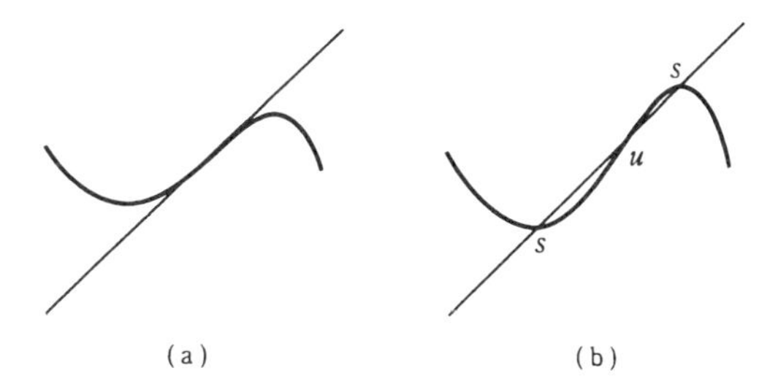

Figure 3.14. (a) The starting point of a pitchfork bifurcation and (b) the pair creation of an unstable (u) and stable (s) fixed points.

where the first equality follows since $f(x_k) = x_k$ and f and g stand for f_R and g_R, respectively. Accordingly the condition $g'''(x_k) < 0$ leads to the third-order contact of $y = g_R(x)$ and $y = x$ at $x = x_k$ as shown in figure 3.14(a). In general, it follows that $g'(x_k) > 1$ for $R > R_n$, which implies two real solutions on both sides of $x = x_k$ as shown in figure 3.14(b). That is, the solution $x = x_k$ becomes unstable and there appear new stable solutions on both sides leading to a pitchfork bifurcation. What is important, therefore, is the inequality $g'''(x_k) < 0$. What has been explained so far may be expressed mathematically as follows. The equation giving the fixed points for $R > R_n$ is

$$g_R(x) - x = \frac{1}{3!}g_R'''(x_k)y^3 + \frac{1}{2!}g_R''(x_k)y^2 + (g_R'(x_k) - 1)y = 0, \tag{3.14}$$

where $y = x - x_k$. It has solutions, besides $y = 0$, given by

$$\frac{1}{3!}g_R'''(x_k)y^2 + \frac{1}{2!}g_R''(x_k)y + g_R'(x_k) - 1 = 0.$$

The discriminant of this equation is

$$D = \frac{1}{4}(g_R''(x_k))^2 - \frac{2}{3}g_R'''(x_k)(g_R'(x_k) - 1)$$

which can be rewritten as

$$D = \frac{1}{4}\left(\frac{\partial g_{R_n}''(x_k)}{\partial R}\right)^2 (\delta R)^2 - \frac{2}{3}g_R'''(x_k)\left(\frac{\partial g_{R_n}'(x_k)}{\partial R}\right)\delta R \tag{3.15}$$

where $R = R_n + \delta R$. Since

$$\frac{\partial g_{R_n}'(x_k)}{\partial R} > 0$$

in most cases, $g'''(x_k) < 0$ implies $D > 0$ leading to two real solutions.

Therefore the necessary condition for the infinite sequence of pitchfork birfurcations is

$$g_R'''(x_k) < 0$$

at $R = R_n$ $(n = 1, , 2, 3, \ldots)$. In fact, this condition is equivalent to negativity of the *Schwarz derivative*, which is called the Schwarz condition. The Schwarz derivative $S[f]$ of a function f is defined by

$$S[f] \equiv \frac{f'''}{f'} - \frac{3}{2}\left(\frac{f''}{f'}\right)^2. \tag{3.16}$$

The Schwarz derivative has the property that $S[f] < 0$ implies $S[f^n] < 0$.

Problem 7. Show

$$S[f^n]_{x=x} = \left(\frac{\mathrm{d}f^{n-1}}{\mathrm{d}x}\right)^2 S[f]_{x=f^{n-1}(x)} + S[f^{n-1}]_{x=x}$$

to prove the above statement.

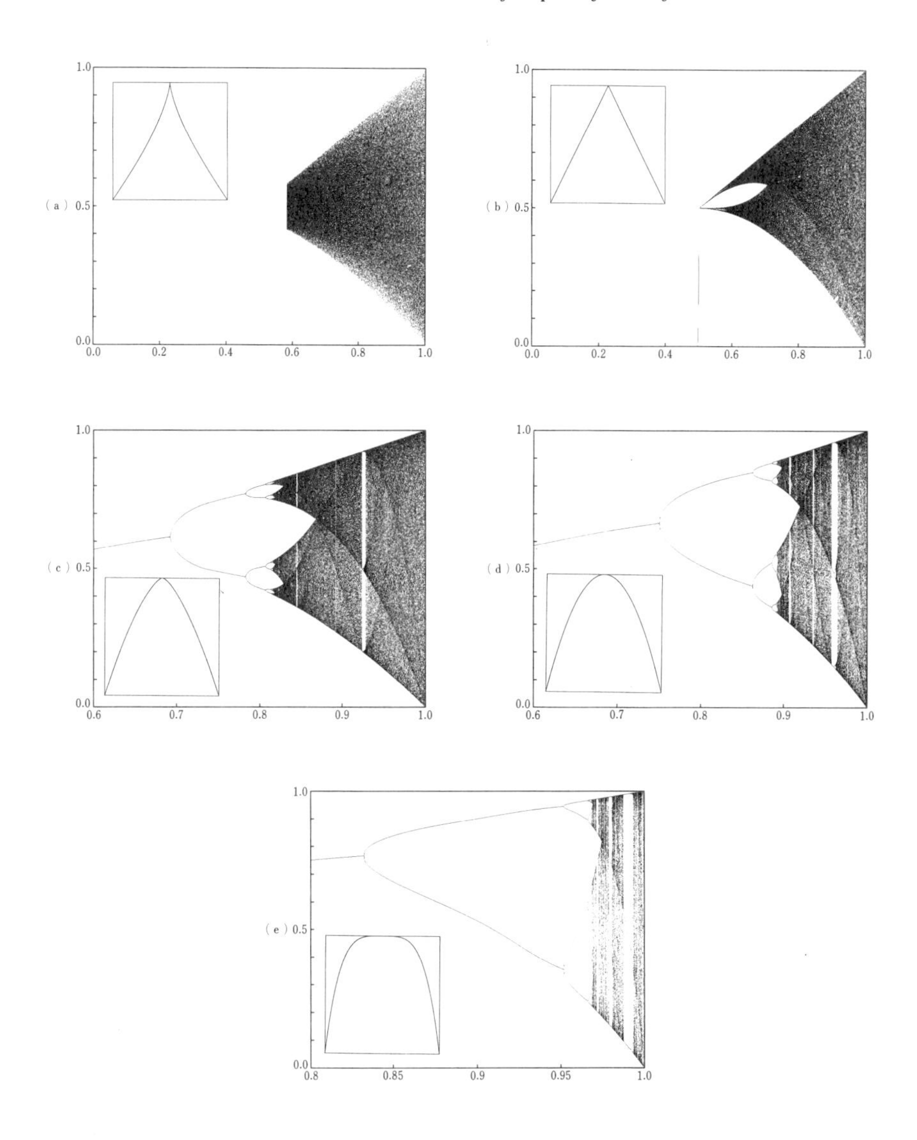

Figure 3.15. Stationary orbits of the map $x_{n+1} = A(1 - |1 - x^{2n}|^p)$. The abscissa denotes the parameter A. (a) $p = 0.7$, (b) $p = 1.0$, (c) $p = 1.5$, (d) $p = 2.0$ and (e) $p = 4.0$.

As mentioned before, f satisfies $f'_{R_n}(x_k) = -1$ and $(f^2_{R_n}(x_k))' = 1$ at a bifurcation point, which leads to

$$S[f^2] = (f^2)'''.$$

If this is combined with the above statement, it is found that an infinite sequence of pitchfork birfurcations exists if

$$S[f] < 0$$

on the domain I of x. Let us take the logistic map $L_R(x)$ as the map f for example. Then one finds $S[f] = -\frac{3}{2x^2} < 0$ from $f''' = 0$ and $f'' = -2R$.

Problem 8. Find the Schwarz derivative of the following functions.

(1) $f(x) = x^n$ (2) $f(x) = A(1 - |1 - 2x|^p)$.

Problem 9. Show that

$$S[f] = -2\sqrt{f'}\frac{d^2}{dx^2}\left(\frac{1}{\sqrt{f'}}\right).$$

The Feigenbaum route has been observed in many one-dimensional maps with smooth peaks. In contrast, even a pitchfork bifurcation has not been observed in maps, such as the tent map, that do not satisfy the *Schwarz condition* $S[f] < 0$. If a map f has a single peak that is not quadratic (i.e. $f''(x_0) = 0$), the ratio δ of the distance of the adjoining bifurcation points defined by the first equality of equation (3.4) is not the Feigenbaum constant. For example, the peak is quartic if $f''''(x_0) \neq 0$, in which case one obtains $\delta = 7.285\ldots$.

Table 3.1. The values of p and δ for $x_{n+1} = A\{-(x - \frac{1}{2})^p + \frac{1}{2^p}\}$.

p	2	4	6	8
δ	4.669	7.285	9.296	10.948

Figure 3.15 shows the map $x_{n+1} = A(1-|1-2x_n|^p)$ for $n \geq 200$. Pitchfork bifurcations and windows are observed for $p > 1$, since the Schwarz derivative is negative (see problem 8) and the map is almost flat in the vicinity of the peak. If one compares the cases $p = 2$ and $p = 4$ together, one finds that the distance between adjoining bifurcation points more suddenly decreases for $p = 4$ as A increases. Table 3.1 shows the ratio δ given by equation (3.4) obtained by numerical computations for $p = 2, 4, 6, 8$. The route to chaos through period doubling phenomena mentioned here is a typical scenario in the genesis of chaos, that is observed in many experiments. Figure 3.16 shows an example of this in a laser experiment. Amplitude modulations of laser light are measured in this experiment while the length of the cavity of a xenon laser is fine tuned. In these figures, the abscissa is the amplitude modulation frequency while the ordinate is the intensity of the light with that frequency in a log scale. The fundamental frequency changes as $f \to f/2 \to f/4$ in figure 3.16(a)–(c) while the period T is doubled as $T \to 2T \to 4T$ since $T = 1/f$.

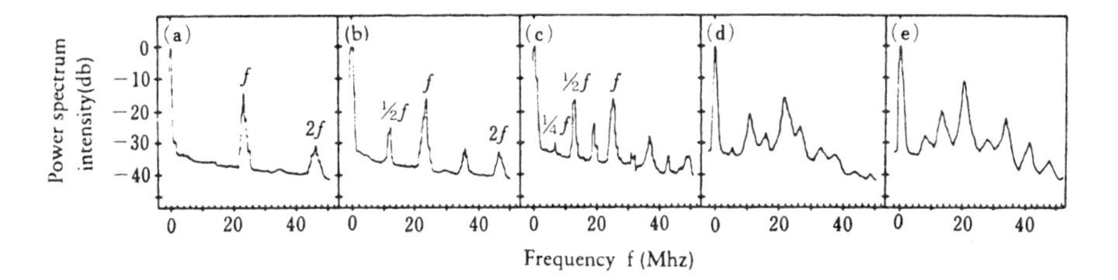

Figure 3.16. (a)–(c) the period doubling phenomena, (d) chaos and (e) the period 3 case observed in laser oscillations. (From Gioggia R S and Abraham N B 1983 *Phys. Rev. Lett.* **22** 650.)

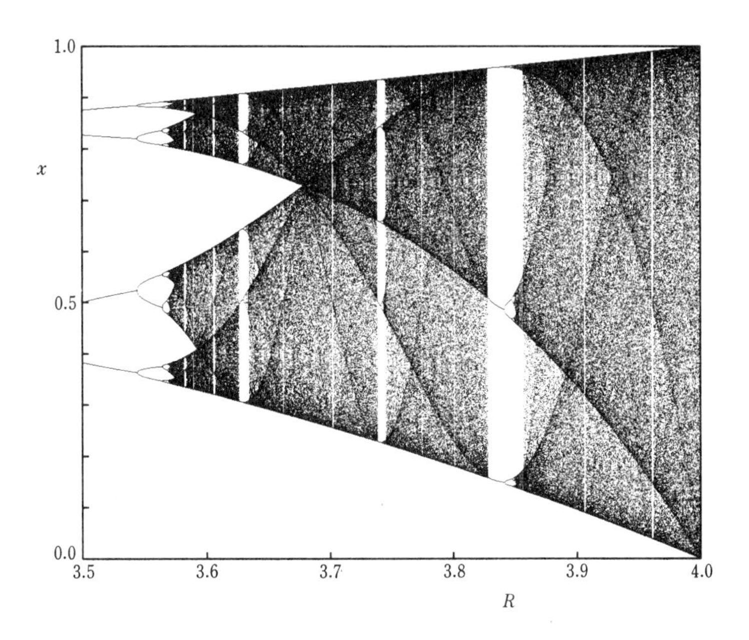

Figure 3.17. Distribution of windows in the logistic map $L_R(x)$.

3.3 Windows

Figure 3.17 shows the distribution of orbits (asymptotic orbits) of the logistic map as a function of R. One finds from this figure that there exist regions in R_∞ $(= 3.569\,9456) < R < 4$, where periodic orbits, without chaotic behaviour, appear. These regions are called the *windows*. The stable orbit in the largest window around $R \sim 3.84$ has period 3 and is called the *period 3 window*. When the parameter R takes a value in a window, an orbit starting with almost every initial condition asymptotically approaches the periodic orbit. Here, 'almost every' means that there are an uncountable number of initial conditions that do not asymptotically approach the periodic orbit in the sense of Li–Yorke in a period 3 window for example. These nonperiodic orbits are not observable in general, since orbits that do not approach the periodic orbit have measure zero.

There exist a (countably) infinite number of windows with a finite width. The period 3 window mentioned above is the unique one with this period. Then how many period 4 windows are there and where are they? The answer is that it is also unique and is found around $R \sim 3.962$. It is also known from numerical computations that if all the widths of the windows are summed up, it amounts to approximately 10% of $4 - R_\infty$.

The window with period 3 is the largest one, which starts at $R = 1+\sqrt{8} = 3.828\,4271\ldots$ as shown in appendix 3A. Figure 3.18(a) shows the map $L_R^3(x)$ obtained by iterating the logistic map three times, from which one finds the map touches the diagonal line at three points α, β and γ. If R is slightly increased from this value $R_c = 1 + \sqrt{8}$, these three degenerate solutions bifurcate and each of them yields two real solutions, one of which is a stable periodic point while the other is an unstable periodic point (see figure 3.18(b)). (In the figure, the symbol + (−) denotes a stable (unstable) periodic point.)

Problem 10. Show that the three solutions α, β and γ are not separately degenerate but simultaneously degenerate.

These stable periodic points attract almost every orbit. This kind of bifurcation, that produces a stable and an unstable periodic point in a pair after touching the diagonal line, is called the *tangent bifurcation*. Not only the

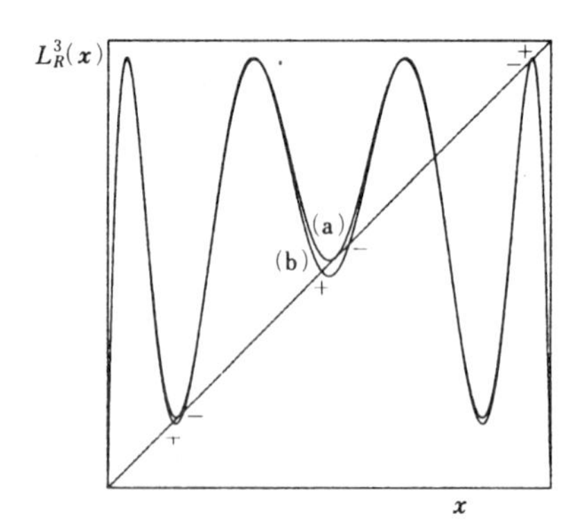

Figure 3.18. The graph of the map $L_R^3(x)$, the three-times-iterated logistic map. (a) The window starts at $R = 1 + \sqrt{8} = 3.828\,4271\ldots$. (b) A stable periodic point (+) and an unstable periodic point (−) in the window ($R = 3.845\,00$).

period 3 window but also windows of any period are produced by the tangent bifurcations. The stable periodic points produced by the bifurcation undergo successive pitchfork bifurcations within the window, leading to 'small chaos' there. Magnification of figure 3.17 shows that this small chaos is obtained from the reduction of the stable orbits of the logistic map on the whole range $0 < R \leq 4$. There exists a window within the small chaos with period 3 (in fact period 9 (=3×3)), that corresponds to the original window with period 3. This small window has smaller chaos, within which there are further smaller windows and so on, thus leading to a marvellous self-similar structure.

A *superstable orbit* is a periodic orbit in a window, which passes through $\frac{1}{2}$. The expansion rate of this orbit under this map vanishes and the Lyapunov number is $-\infty$. Let us consider a superstable orbit taking an example from a window with period 3. The value of R producing a superstable orbit is a solution of $L_R^3\left(\frac{1}{2}\right) = \frac{1}{2}$. This is written explicitly as

$$R^7 - 8R^6 + 16R^5 + 16R^4 - 64R^3 + 128 = 0. \qquad (3.17)$$

Problem 11. Derive equation (3.17).

Equation (3.17) has a factor $(R - 2)$ since there exists a period 1 orbit that passes through the peak when $R = 2$. If equation (3.17) is divived by this factor, one obtains

$$R^6 - 6R^5 + 4R^4 + 24R^3 - 16R^2 - 32R - 64 = 0.$$

The solutions of this equation are not obtained through a simple algebraic method and hence cannot be factorized. However this equation can be easily solved numerically with a calculator or a personal computer by the use of the Newton method. The superstable orbit in the window with period 3 thus obtained has the solution

$$R_S = 3.831\,874\,055\ldots.$$

This number may be employed as a representative value of the position of the window.

The Newton method is quite a powerful tool to solve this kind of higher order polynomial equation and its principle and the computer program are given in appendix 3B.

Next let us estimate the size of the window with period 3 and the extension of the small chaos within it. The reader will obtain a rough idea about the similarity in chaos through this calculation.

The map $L_R^3(x)$ is first expanded aroud $x = \frac{1}{2}$ by putting $x = \frac{1}{2} + x'$. Keeping terms up to the second order in x', one obtains

$$\begin{aligned} L_R^3(x) \sim \frac{R^3}{256}\{&(16R^4 - 96R^3 + 128R^2 + 128R - 256)x'^2 \\ &- R^4 + 8R^3 - 16R^2 - 16R + 64\}. \end{aligned} \qquad (3.18)$$

This amounts to approximating $L_R^3(x)$ near the minimum at $x = \frac{1}{2}$ as a quadratic function. If a superstable value R_s of R is substituted into the above equation, one has the approximation

$$35.63x'^2 + 0.50.$$

Thus the 'amplitude'[1] of this parabola is reduced from that of the original $L_R(x) = -Rx'^2 + \frac{R}{4}$ by approximately $\frac{1}{9}\left(\sim \frac{4}{35.63}\right)$ at $R = 4$. (Note that the similarity ratio of $y = f(x)$ to $ay = f(ax)$ is $1 : \frac{1}{a}$.) Accordingly in the small chaos in a window, the amplitude of the central small chaos is estimated to be roughly $\frac{1}{9} \sim 0.1$, which is in good agreement with the numerical result given in figure 3.19.

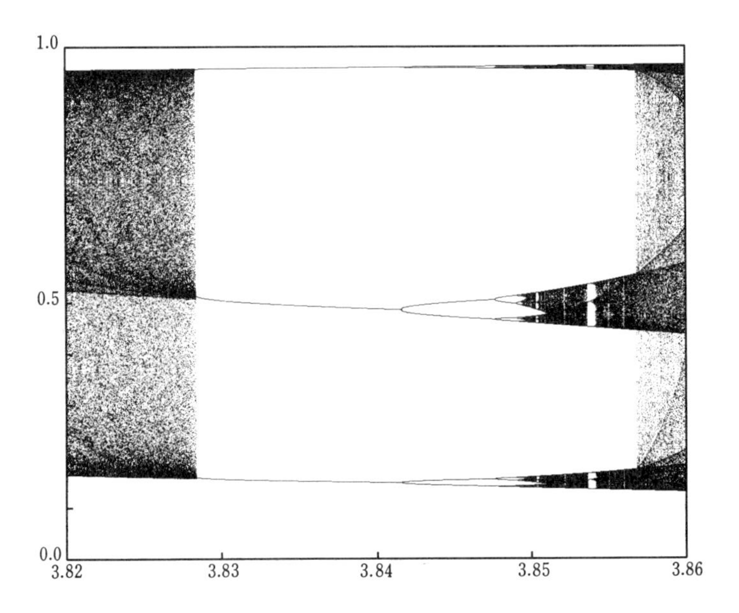

Figure 3.19. The enlargement of stable orbits in the period 3 window corresponding to figure 3.18.

Chaos in this small parabola spreads all over the amplitude of $\frac{1}{9}$ at the point where the window terminates. As the minimum of figure 3.20 is further reduced, chaotic orbits are not confined within this amplitude any more but occupy the whole range. This phenomenon is called the *crisis* (see figure 3.20). The value of R where the crisis takes place (or the window terminates) is estimated from the approximate similarity between the map $L_R(x)$ itself and the behaviour of

[1] In the following, the extension along the R-axis is called the 'width', while that along the x-axis is called the 'amplitude'.

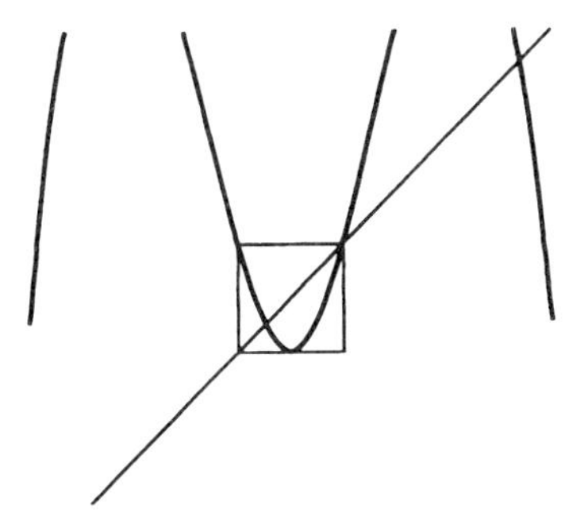

Figure 3.20. $L_R^3(x)$ at the end point of the period 3 window. The peak is pushed out of the frame as R is further increased. Then chaotic orbits occupy the whole range (the crisis) and the window disappears.

$L_R^3(x)$ near $x' = 0$ $(x = \frac{1}{2})$. In other words, the height of the minimal point when the crisis takes place is considered to be $\frac{1}{2} - \frac{1}{9} \times \frac{1}{2} = \frac{4}{9}$ and hence one solves the equation

$$\frac{R^3}{256}(-R^4 + 8R^3 - 16R^2 - 16R + 64) = \frac{4}{9}$$

with the Newton method to find $R_0 = 3.857\,082\,826\ldots$. Thus the width of the window is estimated as $R_0 - R_c = 0.0287$, which is in good agreement with the numerical results.

Generally speaking, when the map near its peak is approximated by a parabola of the form $A(R)x'^2 + B(R)$, the maximum width of the chaotic region is of the order of $4/A(R)$ and the variation ΔR of R as the peak shifts by half of this width is obtained from

$$B'(R)\Delta R = \frac{2}{A(R)}$$

as

$$|\Delta R| = \left|\frac{2}{A(R)B'(R)}\right|$$

which is a rough estimate of the width of the window.

How many windows are there? There is only one window with period 3, namely, the large one around $R \sim 3.84$. What about the positions and the number of period 4 windows? The answer is there is only one around $R \sim 3.963$. How many period 5 windows are there? In fact, there are three of them. Their positions, $R = 3.738\,92\ldots, 3.905\,71\ldots$ and $3.990\,27\ldots$ have been obtained by solving the polynomial equation $L_R^3(\frac{1}{2}) = \frac{1}{2}$ in R by the Newton method (see appendix 3B). The window around the first one occupies a considerable area in figure 3.17, while that around the second one is barely seen and the last one is unrecognizable in this figure. In general, the size of the window shrinks as R approaches 4.

Natural questions one may ask are why three windows with period 5 are distributed in such different locations and what characterizes the windows including their distribution. The answer to both questions is given by the superstable periodic orbits in the windows. This is because the Lyapunov number of a superstable periodic orbit is negative (in fact, $-\infty$) and there exist stable periodic orbits in the vicinity of R giving this superstable orbit. A window extends over the region of R where these stable periodic orbits exist. Figures 3.21(a), (b) and (c) show the superstable periodic orbits in three period 5 windows in increasing order of R. This order is the same as that of the topological entropy associated with each orbit, namely

$$\log 1.512\,88, \qquad \log 1.722\,08, \qquad \log 1.927\,56$$

respectively. The method to evaluate the topological entropy of a superstable periodic orbit will be given in appendix 3C.

The number of windows may be obtained from the number of periodic orbits at $R = 4$. For example, there are $5 \times 2 \times 3 = 30$ periodic orbits, both stable and unstable, generated by tangent bifurcations in three period 5 windows. These periodic orbits are preserved up to $R = 4$ and there are $2^5 - 2 = 30$ periodic orbits with period 5 at $R = 4$ as mentioned in chapter 2. Therefore if the periodic orbits are assumed to persist up to $R = 4$, which is a guaranteed fact for the logistic map, the number of windows may be computable from the number of periodic orbits at $R = 4$.

A periodic orbit with period p cannot be generated by a pitchfork bifurcation, when p is odd, but can be generated only by a tangent bifurcation. When p is a prime number, in particular, the number of the periodic orbits is $2^p - 2$ [2] while the number of windows is

$$\frac{2^p - 2}{2p} = \frac{2^{p-1} - 1}{p}.$$

(The prime number p (≥ 3) divides $2^{p-1} - 1$ by Fermat's theorem.)

Problem 12. Find the number of windows with $p = 3, 5, 11$ and 23.

Roughly speaking, the number of windows with period k is $\frac{2^{k-1}}{k}$. If one takes $k = 50$, for example, one obtains an astronomically large number 1.12×10^{13}.

Finally let us consider the condition for a map to have windows. The answer is found from our discussions on windows so far. That is, a window shrinks if the curvature at the peak of the local 'parabola' leading to a tangent

[2] It is easy to see from figure 2.5(b) that there are 2^p crossing points of $L_R^p(x)$ with the diagonal line. When p is prime, two of the 2^p points are the solutions of $L_R(x) = x$ while all the other points cannot be solutions of $L_R^q(x) = x$ with $q < p$ since no number divides p except 1 and p.

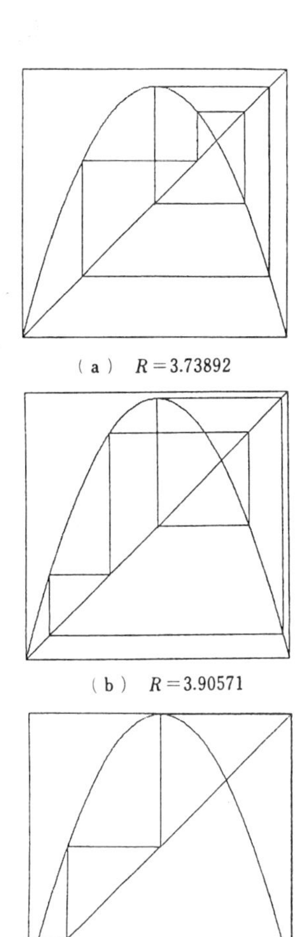

Figure 3.21. Superstable orbits in period 5 windows. The parameter increases in the order of (a), (b) and (c). This order is also the order of increasing topological entropy.

bifurcation is large. As a limiting case of this observation, the tent map has no windows. In other words, the window is found over a region of R around superstable periodic orbits, where the Lyapunov number is negative. For this region to exist, therefore, the map must have a smooth peak as the logistic map.

It should be noted that our discussions on the properties of the windows raise the following question. That is, there seem to be only periodic orbits and no chaos in windows. However, there is a period 3 window with periodic orbits and hence there must be chaos (the scrambled set) in the sense of Li–Yorke. What is this unobserved chaos? The answer is there are an uncountable number of chaotic orbits with measure zero, that cannot be found in numerical computations. A well known example of this kind of set is the Cantor set. In fact, if one starts from an initial value that is not a periodic point in a window, one finds that the point behaves irregularly and eventually falls into the periodic orbit.

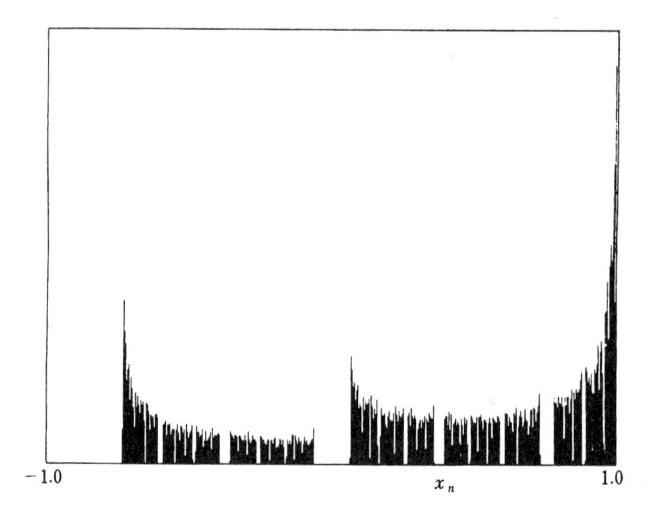

Figure 3.22. The invariant measure of the logistic map in the period 3 window. The map is unstable as a whole and this measure has been obtained by the inverse map. The map employed in this computation is $x_{n-1} = 1 - ax_n^2$ with $a = 1.754\,877\,67$. If this map is transformed into a map of the form $x_{n+1} = Rx_n(1 - x_n)$, the parameter R becomes $3.831\,87\ldots$, the same number employed in figure 3.23. (From Kantz H and Grassberger P 1985 *Physica* D **17** 75.)

The number of repetitions the point makes before it falls into the periodic orbit heavily depends on the initial point. Although numerical error should be taken into account in this attempt using a computer, this phenomenon suggests that there exists a complex structure in a window. Figure 3.22 shows the invariant measure of the chaos obtained by the inverse map at $R = 3.831\,8740\ldots$, where a superstable periodic orbit exists in the period 3 window. Such chaos is unstable and leaves its trace in the transient phenomena that the system shows while falling into a stable period 3 orbit as shown in figure 3.23.

3.4 Intermittent chaos

Intermittent chaos is observed when the parameter R is slightly smaller than the value at which a window begins. By making R smaller than the parameter range corresponding to the window, one may follow the tangent bifurcation in the reverse direction. Let us put $R = 1 + \sqrt{8} - \delta$ to see the behaviour of the map immediately below $R = R_c = 1 + \sqrt{8}$, the beginning of the period 3 window. Figure 3.24 shows an example of chaos obtained when $\delta = 0.0001$, from which one finds that an orbit is composed of a periodic part with period 3 and a part scattered over the whole region. The former is called *laminar*, while

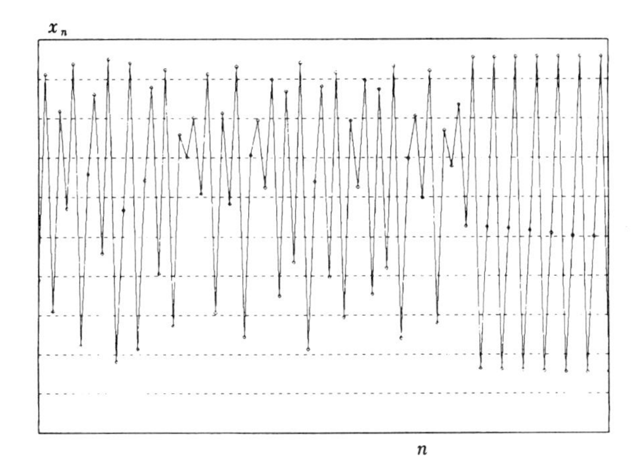

Figure 3.23. A transient irregular orbit in the period 3 window ($R = 3.831\,874\,055$, $x_0 = 0.39$). The length of the irregular part greatly changes as x_0 is varied.

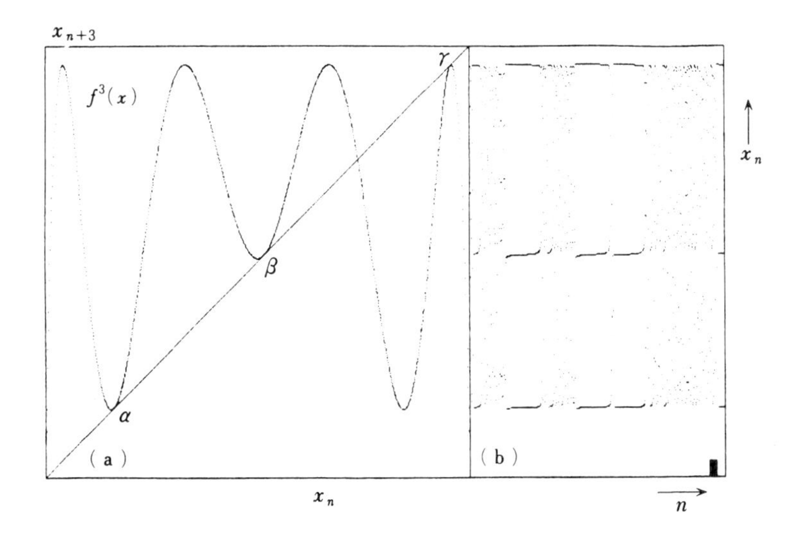

Figure 3.24. Intermittent chaos observed immediately below the period 3 window. Here $R = 1 + \sqrt{8} - \delta$ with $\delta = 0.0001$.

the latter *burst*. (The word 'laminar' originates from laminar flow, meaning a stationary part, while variations in the 'burst' part are explosive.) The laminar part is produced when a part of the map is close to the diagonal line $x_{n+1} = x_n$

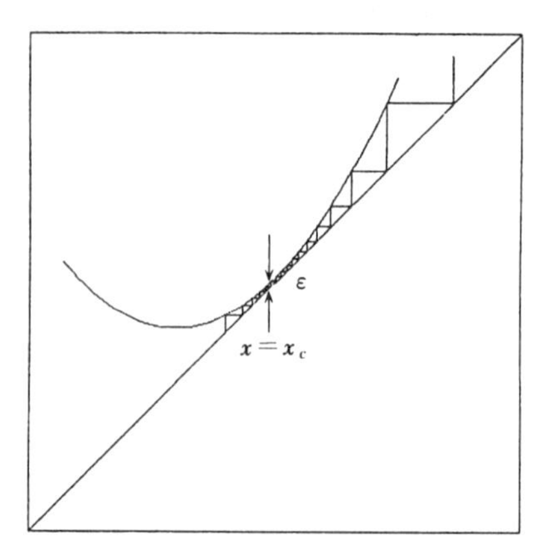

Figure 3.25. An orbit in the laminar part.

so that it takes quite a long time before an orbit passes through the region while keeping the value of x_n almost constant under many iterations (see figure 3.25). The burst is produced when an orbit escapes from the laminar part and spreads into the whole region. The duration of the laminar part becomes longer, of course, as δ is decreased, while it becomes shorter and eventually disappears (to become an ordinary chaos) as δ is increased. This behaviour is depicted in figure 3.26.

Now let us consider how the averaged duration of the laminar part changes with δ. Let us consider a general case first. The map is approximated by a quadratic function around x_c as

$$x_{n+1} = a(x_n - x_c)^2 + x_n + \varepsilon \tag{3.19}$$

when it is separated from the diagonal line ($x_{n+1} = x_n$) by ε at x_c as shown in figure 3.25.

If x_c is subtracted from both sides and $x_n - x_c$ is again written as x_n, this approximation becomes

$$x_{n+1} = ax_n^2 + x_n + \varepsilon. \tag{3.20}$$

Here x_n hardly changes in the laminar part. This equation is Taylor expanded as

$$x_{n+1} = x_n + \frac{\mathrm{d}}{\mathrm{d}n}x_n + \frac{1}{2!}\frac{\mathrm{d}^2}{\mathrm{d}n^2}x_n + \ldots$$

by regarding n as a continuous parameter.

Since the variation of x_n is mild in the laminar part, higher derivatives may be dropped to yield

$$x_{n+1} - x_n \simeq \frac{\mathrm{d}x_n}{\mathrm{d}n}.$$

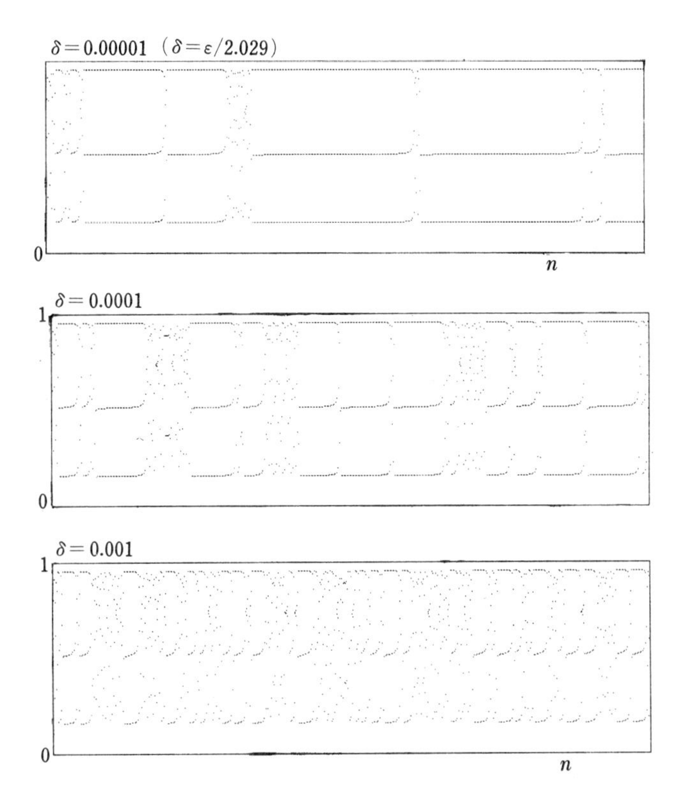

Figure 3.26. Intermittent chaos when δ is varied.

By suppressing the subscript n, one arrives at the equation

$$\frac{\mathrm{d}x}{\mathrm{d}n} = ax^2 + \varepsilon. \tag{3.21}$$

This equation is integrated from $n = n_1$ to n_2 to yield

$$n_2 - n_1 = \frac{1}{\sqrt{\varepsilon a}} \left\{ \tan^{-1}\left(\frac{x_2}{\sqrt{\varepsilon/a}}\right) - \tan^{-1}\left(\frac{x_1}{\sqrt{\varepsilon/a}}\right) \right\} \tag{3.22}$$

where $x_1 = x_{n_1}$ and $x_2 = x_{n_2}$.

Problem 13. Derive equation (3.22) from equation (3.21).

In equation (3.22), x_1 is the beginning point of the laminar orbit. Provided that the distribution of x_1 is symmetric around x_c, i.e. $x = 0$, the duration $\langle l \rangle$

of the laminar part averaged over x_1 is given by

$$\langle l \rangle = \frac{1}{\sqrt{\varepsilon a}} \tan^{-1}\left(\frac{x_2}{\sqrt{\varepsilon / a}}\right) \tag{3.23}$$

since $\tan^{-1}$ is an odd function. If ε is sufficiently small, this simplifies as

$$\langle l \rangle = \frac{1}{\sqrt{\varepsilon a}} \frac{\pi}{2}. \tag{3.24}$$

One needs to estimate ε and a to apply equations (3.23) and (3.24) to the intermittent chaos just below the period 3 window.

The central minimal point in the third order map $L_R^3(x)$ (the point β in figure 3.24) may be taken as x_c in equation (3.19). One may use equation (3.18) as the approximation of $L_R^3(x)$ around $x = \frac{1}{2}$ since this x_c is close to $x = \frac{1}{2}$. Let us write equation (3.18) as

$$L_R^3(x) = A(R)\left(x - \frac{1}{2}\right)^2 + B(R). \tag{3.25}$$

This is also rewritten as

$$L_R^3(x) = A(R)\left\{x - \left(\frac{1}{2} + \frac{1}{2A(R)}\right)\right\}^2 + x + \left\{B(R) - \frac{1}{4A(R)} - \frac{1}{2}\right\} \tag{3.26}$$

which determines x_c in equation (3.19) as

$$x_c = \frac{1}{2} + \frac{1}{2A(R_c)} = 0.5142\ldots.$$

This value is in good agreement with the exact one $x_c = 0.5143\ldots$ derived in appendix 3A. The coefficient of the quadratic part is

$$A(R_c) = 35.2 \tag{3.27}$$

while the distance ε between the parabola and the diagonal line $(x_{n+1} = x_n)$ is estimated as

$$\begin{aligned}\varepsilon(R) &= \varepsilon(R_c) + \frac{d\varepsilon(R_c)}{dR}(R - R_c) + \ldots \\ &= \frac{d}{dR}\left(B(R) - \frac{1}{4A(R)} - \frac{1}{2}\right)\bigg|_{R=R_c}(R - R_c) + \ldots \\ &\simeq 2.03\delta.\end{aligned} \tag{3.28}$$

If one is more careful in deriving equations (3.27) and (3.28), one notices that the value $A(R_c)$ given by equation (3.27) should have more dependence on the position since it is the second derivative of the curve. The coefficient of the

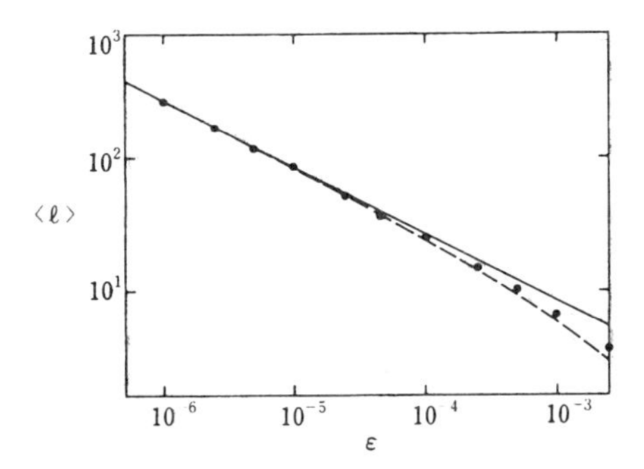

Figure 3.27. The average length of the laminar part in the intermittent chaos as a function of the control parameter ε. The dots are the result of the numerical computation while the solid (broken) line is calculated with equation (3.34) (equation (3.23)). (From Hirsh J E, Huberman B A and Scalapino D J 1981 *Phys. Rev.* A **25** 519.)

second order term centred at x_c is derived from equations (3A.1) and (3A.2) in appendix 3A as

$$A = R^6\beta(R\beta + 1 - R)(\beta - \alpha)^2(\beta - \gamma)^2 \tag{3.29}$$

where α, β and γ are points where $L_R^3(x)$ touches the diagonal line ($x_{n+1} = x_n$), the values of which are given in appendix 3A. Substituting $R = R_c = 1 + \sqrt{8}, \alpha, \beta$ and γ into the above equation, one obtains

$$A = 34.1453\ldots. \tag{3.30}$$

The coefficient of δ in equation (3.28) remains unchanged if $\varepsilon(R)$ is estimated with a parabola centred at $x = x_c$ ($= \beta$). The expression thus obtained is

$$\begin{aligned}\frac{\varepsilon}{\delta} &= -2R\alpha\gamma + 3\alpha + R\gamma - \frac{7\beta}{R}\\ &= 2.029\,45\ldots.\end{aligned} \tag{3.31}$$

From equations (3.30) and (3.31), one finds

$$A\frac{\varepsilon}{\delta} = 68.2964\ldots \tag{3.32}$$

is just what one needs to evaluate the average length $\langle l \rangle$ of the laminar part

$$\langle l \rangle = \frac{\pi}{2\sqrt{A\frac{\varepsilon}{\delta}}}\frac{1}{\sqrt{\delta}} \tag{3.33}$$

in terms of δ, the deviation of R from R_c.

This value is the same, of course, irrespective of the choices of x_c from α, β and γ. The explicit value is

$$\langle l \rangle = \frac{0.190}{\sqrt{\delta}} = \frac{0.270}{\sqrt{\varepsilon}} \qquad (x_c = \beta) \tag{3.34}$$

which is in excellent agreement with the numerical result shown in figure 3.27.

Problem 14. Derive equations (3.29) and (3.30) by making reference to appendix 3A.

Chapter 4

Chaos in realistic systems

Chaos has been explained taking one-dimensional systems as an example so far. Undoubtedly various aspects of chaos can be explained clearly if one-dimensional maps are considered. However, a natural question to ask is whether these quite simple equations are meaningful in the description of natural phenomena. Chaos in a system of differential equations and chaos in a realistic system are considered in the present chapter to answer this question.

4.1 Conservative system and dissipative system

Intuitively speaking, a conservative system is something like an oscillator without friction. There is no energy dissipation in this case and the oscillator executes a periodic motion forever. A nonconservative system is called a dissipative system, a typical example of which is a damped oscillation. If the energy of a whole system does not dissipate this system is called a conservative system, while if the energy increases or decreases with time it is called a dissipative system as seen in the above examples.

Let us consider a simple harmonic oscillator with mass m and spring constant k as an example. The equation of motion is

$$m\frac{\mathrm{d}^2x}{\mathrm{d}t^2} = -kx \tag{4.1}$$

whose general solution is

$$x = A\cos\omega_0 t + B\sin\omega_0 t \tag{4.2}$$

where $k/m = \omega_0^2$.

If one puts $m\dot{x} = y$ in equation (4.1), where the dot represents the time derivative, one obtains

$$\begin{cases} \dot{x} = \dfrac{1}{m}y \\ \dot{y} = -kx. \end{cases} \tag{4.3}$$

Physically y is the momentum of the oscillator. From these equations, one obtains

$$\frac{\mathrm{d}y}{\mathrm{d}x} = -mk\frac{x}{y} \tag{4.4}$$

by making use of the formula $\frac{\mathrm{d}y}{\mathrm{d}t} \Big/ \frac{\mathrm{d}x}{\mathrm{d}t} = \frac{\mathrm{d}y}{\mathrm{d}x}$ for the derivative of parametrized functions. This differential equation is integrated to yield

$$\frac{1}{2m}y^2 + \frac{1}{2}kx^2 = E. \tag{4.5}$$

The parameter E is the integration constant with the dimension of energy and equation (4.5) indicates that the sum of the kinetic energy and the potential energy remains constant.

This relation, when expressed in the xy-plane, defines an ellipse. The xy-plane is called the phase plane while the ellipse is called the orbit.[1]

Problem 1. Show that the energy E and the constants A and B in equation (4.2) satisfy

$$E = \frac{k}{2}(A^2 + B^2).$$

Problem 2. Show that the area of the ellipse S and the energy E are related by the relation

$$E = V_0 S$$

where $V_0 = \omega_0/2\pi$. The size of the ellipse increases with E, although the period T of the oscillator remains constant.

An orbit that passes the point (x_0, y_0) at a time t_0 can be specified in the phase plane by choosing A and B properly. In our case, this can be achieved by solving

$$\begin{pmatrix} x_0 \\ y_0 \end{pmatrix} = \begin{pmatrix} \cos\omega_0 t_0 & \sin\omega_0 t_0 \\ -\omega_0 \sin\omega_0 t_0 & \omega_0 \cos\omega_0 t_0 \end{pmatrix} \begin{pmatrix} A \\ B \end{pmatrix} \tag{4.6}$$

in favour of A and B. The pair (A, B) corresponding to a given (x_0, y_0) is uniquely found since the matrix above is regular.

[1] More generally, let us consider a system of equations

$$\frac{\mathrm{d}x_i}{\mathrm{d}t} = f_i(x_1, x_2, \ldots, x_n) \qquad (i = 1, 2, \ldots, n)$$

which has no explicit t-dependence in the right hand side. Such a system is called an autonomous system. The space $(x_1, x_2, \ldots, x_n)$ is called the *phase space*. The solution to this system of equations has n integration constants in general, which are fixed by the initial condition at $t = 0$. Thus the solution defines a curve in the phase space, which is called the *phase orbit* or simply the orbit. Note that the phase space of a Hamilton system, such as a simple harmonic oscillator, is always even dimensional.

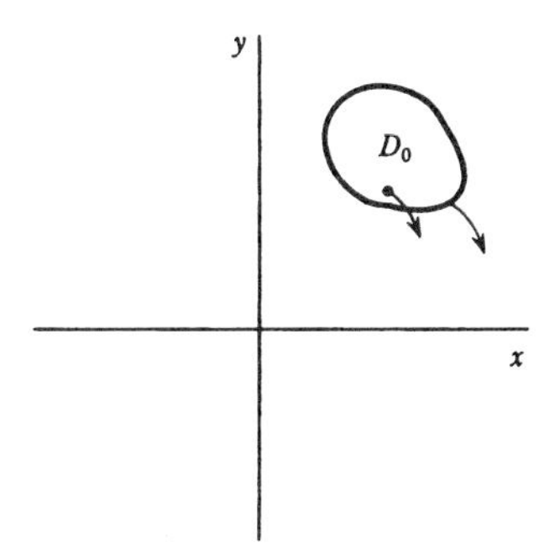

Figure 4.1. The motion of a point along an orbit and that of the domain D_0 in the phase plane.

Next, consider a general system of equations in the two-dimensional phase plane,

$$\begin{aligned} \dot{x} &= f(x, y) \\ \dot{y} &= g(x, y). \end{aligned} \tag{4.7}$$

Let

$$\begin{aligned} x &= x(A, B; t) \\ y &= y(A, B; t) \end{aligned} \tag{4.8}$$

be the solution of the above system, specified by the constants A and B. Consider, as the set of initial conditions, a domain D_0 in the xy-plane shown in figure 4.1. The area S of D_0 is written as

$$S = \int\int_{D_0} \mathrm{d}x\mathrm{d}y. \tag{4.9}$$

Since a point (x, y) in D_0 at a time t corresponds uniquely to a pair of constants (A, B), S can be rewritten as

$$S(t) = \int\int_{\Delta_0} \frac{\partial(x, y)}{\partial(A, B)} \mathrm{d}A\mathrm{d}B \tag{4.10}$$

where

$$\frac{\partial(x, y)}{\partial(A, B)} = \begin{vmatrix} \dfrac{\partial x}{\partial A} & \dfrac{\partial x}{\partial B} \\ \dfrac{\partial y}{\partial A} & \dfrac{\partial y}{\partial B} \end{vmatrix}$$

is the determinant of the Jacobian and Δ_0 is the domain in the AB-plane corresponding to D_0.

Next we consider how $S(t)$ changes with time. Let us analyse a system of linear differential equations as a special case. The general solution of a second-order differential equation is expressed as

$$x = Ax_1 + Bx_2$$

where x_1 and x_2 are the fundamental solutions. Then the momentum y is given by

$$y = \dot{x} = A\dot{x}_1 + B\dot{x}_2.$$

Accordingly, the Jacobian is

$$\frac{\partial(x, y)}{\partial(A, B)} = \begin{vmatrix} x_1 & x_2 \\ \dot{x}_1 & \dot{x}_2 \end{vmatrix} \equiv W(x_1, x_2) \tag{4.11}$$

where $W(x_1, x_2)$ is called the Wronski determinant or the Wronskian.

Thus, the time dependence of $S(t)$ is determined by $W(x_1, x_2)$:

$$\frac{\mathrm{d}}{\mathrm{d}t} W(x_1, x_2) = x_1\ddot{x}_2 - x_2\ddot{x}_1. \tag{4.11$'$}$$

For a simple harmonic oscillator, for example, one finds $\frac{\mathrm{d}W}{\mathrm{d}t} = 0$ since $\ddot{x}_1 = -\omega_0^2 x_1$ and $\ddot{x}_2 = -\omega_0^2 x_2$. Therefore W is independent of time and the area S in the phase plane remains constant (see figure 4.2).

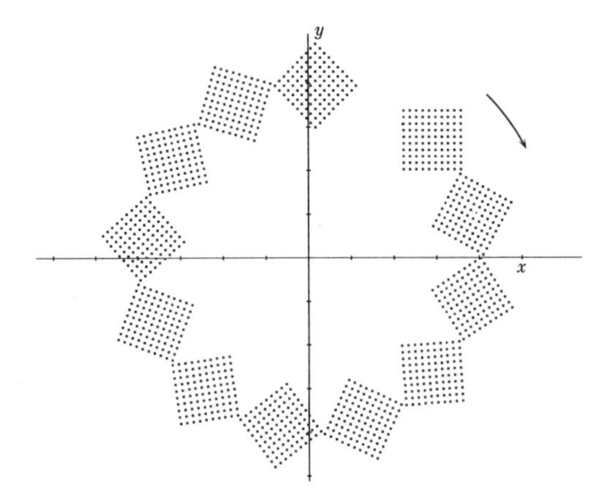

Figure 4.2. The motion of a domain in the phase plane of a simple harmonic oscillator. The area of the domain remains constant. This figure is obtained by calculating the time dependence of $10 \times 10 = 100$ points.

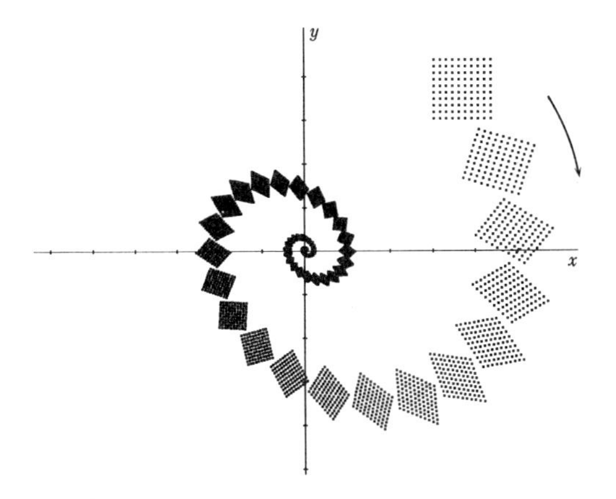

Figure 4.3. The motion of a domain in the phase plane of a damped oscillator. The area decreases exponentially with time and the attractor is the origin.

Problem 3. Consider a damped oscillator with the equation of motion

$$\ddot{x} + 2r\dot{x} + \omega_0^2 x = 0 \qquad (r > 0). \tag{4.12}$$

Show that $W = W_0\,\mathrm{e}^{-2rt}$ and hence the area changes as $S(t) = S(0)\,\mathrm{e}^{-2rt}$ as shown in figure 4.3.

Let us go back to equation (4.10) and consider general cases including nonlinear systems. The time derivative of equation (4.10) yields

$$\frac{\mathrm{d}S(t)}{\mathrm{d}t} = \int\int_{\Delta_0} \left(\frac{\partial(\dot{x}, y)}{\partial(A, B)} + \frac{\partial(x, \dot{y})}{\partial(A, B)} \right) \mathrm{d}A\mathrm{d}B. \tag{4.13}$$

If one substitutes

$$\frac{\partial \dot{x}}{\partial A} = \frac{\partial \dot{x}}{\partial x}\frac{\partial x}{\partial A} + \frac{\partial \dot{x}}{\partial y}\frac{\partial y}{\partial A} \qquad \frac{\partial \dot{x}}{\partial B} = \frac{\partial \dot{x}}{\partial x}\frac{\partial x}{\partial B} + \frac{\partial \dot{x}}{\partial y}\frac{\partial y}{\partial B}$$

into

$$\frac{\partial(\dot{x}, y)}{\partial(A, B)} = \begin{vmatrix} \dfrac{\partial \dot{x}}{\partial A} & \dfrac{\partial \dot{x}}{\partial B} \\[2mm] \dfrac{\partial y}{\partial A} & \dfrac{\partial y}{\partial B} \end{vmatrix}$$

one obtains

$$\frac{\partial(\dot{x}, y)}{\partial(A, B)} = \frac{\partial \dot{x}}{\partial x}\frac{\partial(x, y)}{\partial(A, B)}.$$

Similarly one finds

$$\frac{\partial(x, \dot{y})}{\partial(A, B)} = \frac{\partial \dot{y}}{\partial y}\frac{\partial(x, y)}{\partial(A, B)}.$$

Thus equation (4.13) becomes

$$\frac{\mathrm{d}S(t)}{\mathrm{d}t} = \int\int_{\Delta_0} \left(\frac{\partial \dot{x}}{\partial x} + \frac{\partial \dot{y}}{\partial y} \right) \frac{\partial(x, y)}{\partial(A, B)} \mathrm{d}A\mathrm{d}B. \tag{4.14}$$

If one considers a sufficiently small domain Δ_0 in equation (4.14), it follows from equation (4.10) that

$$\frac{\dot{S}(t)}{S(t)} = \frac{\partial \dot{x}}{\partial x} + \frac{\partial \dot{y}}{\partial y}. \tag{4.15}$$

This expression is very useful since it takes a very simple form. In other words, the variables $\dot{x}, x, \dot{y}$ and y are considered as functions of A and B that are specified by the initial condition and it follows that

$$\frac{\partial \dot{x}_i}{\partial x_i} = \frac{\partial \dot{x}_i}{\partial A}\frac{\partial A}{\partial x_i} + \frac{\partial \dot{x}_i}{\partial B}\frac{\partial B}{\partial x_i} \quad \left(\begin{array}{l} i = 1, 2 \\ x_1 = x, \quad x_2 = y \end{array} \right). \tag{4.16}$$

Equation (4.15) shows that the partial derivatives are evaluated directly from the original equation (4.7). In a damped oscillator considered in problem 3, for example, $y = \dot{x}$ and $\dot{y} = -2ry - x$ yield $\dot{S}(t)/S(t) = -2r$.

Problem 4. Use the solution $x = A\cos\omega t + B\sin\omega t$ and $y = \dot{x}$ in equations (4.15) and (4.16) to show $\dot{S}(t) = 0$.

Equation (4.15) can be extended to the variation of a volume V in the n-dimensional phase space $(x_1, x_2, \ldots, x_n)$, which results in

$$\frac{\dot{V}(t)}{V(t)} = \sum_{i=1}^{n} \frac{\partial \dot{x}_i}{\partial x_i}. \tag{4.17}$$

Let us consider a Hamilton dynamical system with the Hamiltonian H. The canonical equations of motion are

$$\begin{aligned} \dot{p}_i &= -\frac{\partial H}{\partial q_i} \\ & \qquad\qquad (i = 1, 2, \cdots, n) \\ \dot{q}_i &= \frac{\partial H}{\partial p_i} \end{aligned} \tag{4.18}$$

and a volume of this system in the phase space satisfies

$$\frac{\dot{V}(t)}{V(t)} = \sum_{i=1}^{n} \left(\frac{\partial \dot{q}_i}{\partial q} + \frac{\partial \dot{p}_i}{\partial p_i} \right) = \sum_{i=1}^{n} \left(\frac{\partial^2 H}{\partial q_i \partial p_i} - \frac{\partial H}{\partial p_i \partial q_i} \right) = 0.$$

Therefore a volume remains constant in time.

Thus a volume $V(t)$ of a domain in the phase space in a Hamilton dynamical system does not change even when it moves to $V(t')$. Thus we have proved the *Liouville theorem*. The t-independence of the area in the phase plane of a simple harmonic oscillator is an example of this theorem, which has been shown graphically in figure 4.2.

The Hamiltonian H of a Hamilton dynamical system is usually regarded as the energy of the system. For a harmonic oscillator this is given by

$$H = \frac{1}{2m}y^2 + \frac{1}{2}kx^2.$$

A conservative system and a dissipative system are defined, in general, according to whether the volume in the phase space remains constant or not. That is to say,

'a system whose volume in the phase space is constant is called a conservative system while it is called a dissipative sytem otherwise.'

4.2 Attractor and Poincaré section

An orbit of a damped harmonic oscillator is expressed as a solution to equation (4.12), which takes the form

$$x = A\,\mathrm{e}^{-rt} \sin\left(\sqrt{\omega_0^2 - r^2}\,t + \delta\right)$$

and asymptotically approaches the origin as $T \to \infty$. The origin in the present case attracts all the orbits and is called the *attractor.*

It is known when the phase space is two dimensional that there are attracting points as above and attracting closed curves called limit cycles as the set which the orbits approach as $t \to \infty$. Let us consider the attractor of the van der Pol equation

$$\ddot{x} - \varepsilon(1 - x^2)\dot{x} + x = 0 \tag{4.19}$$

as an example of such attracting closed curves. Figure 4.4 shows the orbits and the attractor obtained by numerical computations. The solution of the van der Pol equation at sufficiently later time is just the motion of a point moving on the closed curve. This motion is an oscillation with a fixed amplitude. The van der Pol equation is an example of autonomous oscillations. Such an attractor is called a *limit cycle.*

It should be added that an area in the phase space decreases with time and eventually approaches zero in such systems as a damped oscillator or a van der Pol equation.

Points and closed curves are the only attractors in the two dimensional phase space. The attractors are zero and one dimensional in these examples. As the dimension of the phase space increases, there appear more complex attractors.

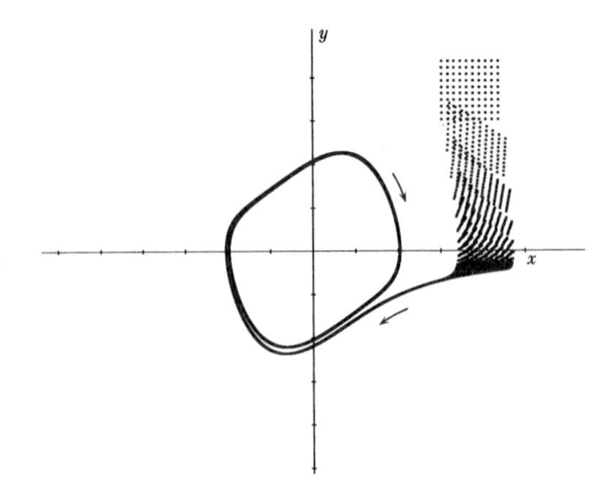

Figure 4.4. The van der Pol equation. Motion of a region in the phase plane and the closed-curve attractor (limit cycle).

Two simple examples with chaotic behaviour are the Rössler model

$$\begin{cases} \dot{x} = -(y + z) \\ \dot{y} = x + \frac{1}{5}y \\ \dot{z} = \frac{1}{5} + z(x - \mu) \end{cases} \tag{4.20}$$

and the Lorenz model

$$\begin{cases} \dot{x} = -\sigma(x - y) \\ \dot{y} = -xz + \gamma x - y \\ \dot{z} = xy - bz \end{cases} \tag{4.21}$$

where μ, σ, γ and b are constants.

These equations are obtained by simplifying equations that describe oscillations in chemical reations or convections and are dissipative. Although the solution is represented as an orbit in a three dimensional phase space as before, the time dependence of the solution is quite complicated compared to those of a damped harmonic oscillation or an autonomous oscillation (the van der Pol oscillator) studied so far. Detailed study of the Rössler equations, for example, reveals that the orbits in the phase space are localized in a flat, ribbon-like region, the whole of which is an attractor (figure 4.5). This attractor is called the *strange attractor*.

Figure 4.6 shows an aperiodic oscillation of x as a function of time. Whether this aperiodic oscillation is chaotic or not is judged according to whether

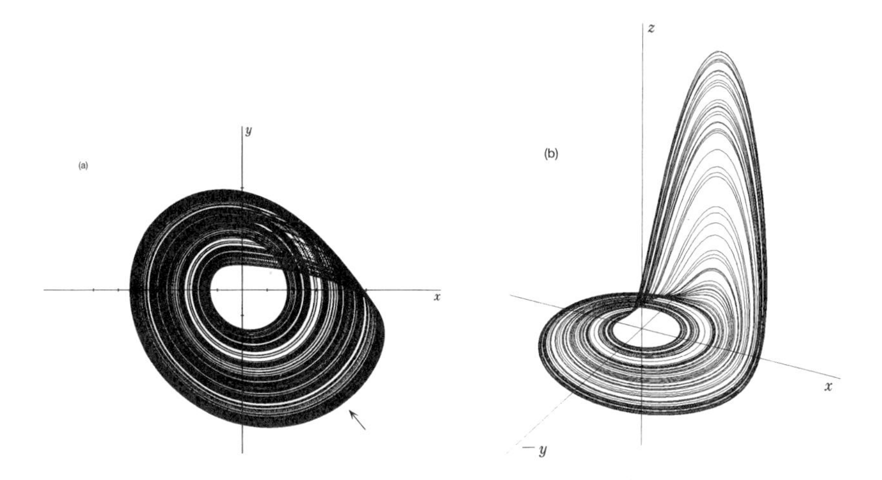

Figure 4.5. (a) The strange attractor of the Rössler model. The whole ribbon is an attractor. Three-dimensional view seen from the arrow is shown in (b). These pictures are computed by putting $\mu = 5.7$ in equation (4.20).

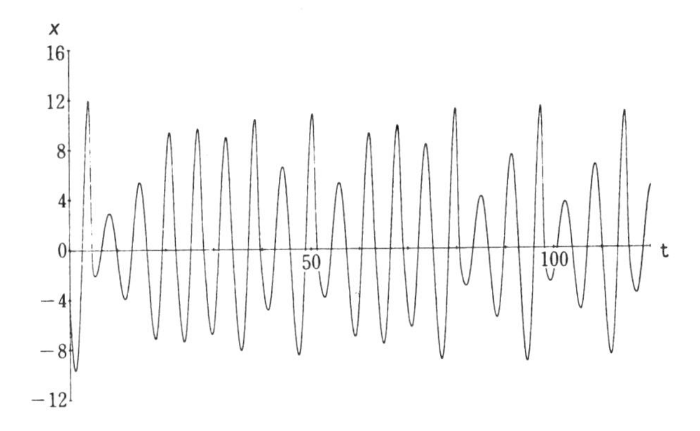

Figure 4.6. An irregular solution of the Rössler model. The time dependence of x in equation (4.20) is shown here.

the largest Lyapunov number is positive or not, which will be elaborated in detail later.

Not only x, but y and z change in aperiodic way. How does an orbit move on the attractor in the phase space when the solution as a whole is aperiodic, as in the present case? Let us consider this problem next.

We employ the *Poincaré section* as the method of analysis. We first mention

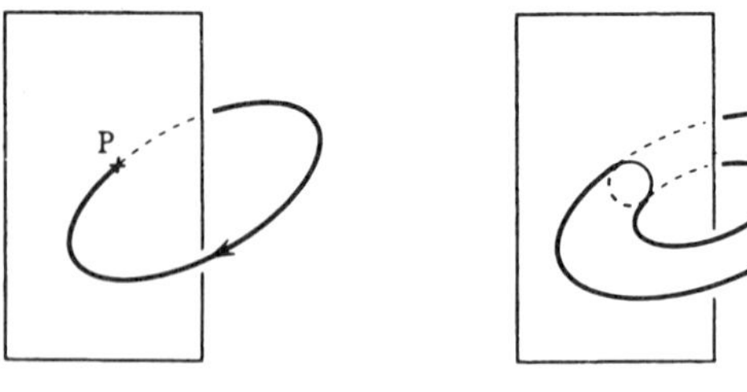

A periodic orbit and a point in the Poincare section.

An orbit on the torus and its Poincare section.

Figure 4.7. An orbit and points in the Poincaré section.

how to construct the Poincaré section and analyse this, as a preliminary to our study of orbits on the strange attractor of the Rössler model.

Let us consider, as the simplest attractor, a simple closed curve in a three-dimensional phase space. We introduce a plane which intersects with this closed curve and call the intersection the Poincaré section. The intersection of this closed curve with the plane is a single point P shown in figure 4.7. The choice of this plane being far from unique, it is possible to choose an appropriate half-plane so that it intersects with the closed curve at a single point. If similar planes are introduced in the analysis of a double-periodic motion and a quasi-periodic motion, their Poincaré sections are two points and a closed curve, respectively. In this way, dimensions of orbits are reduced, by constructing the Poincaré section, in such a way that a closed curve in a three-dimensional space yields a point while a two-dimensional torus yields a closed curve.

Now let us turn to the Poincaré section of the Rössler model. The attractor here is much complicated compared to the examples above: if Poincaré sections are constructed around the z-axis as shown in figure 4.8, each section is found to be a single curve. These lines are stretched and then folded as one makes a complete turn around the z-axis, whose behaviour is essentially same as that of the one-dimensional map mentioned before. Therefore one might expect that a one-dimensional map may be extracted from these Poincaré sections. Let us plot the pair (r_n, r_{n+1}), where r_n (r_{n+1}) is the position of the nth $((n+1)$th$)$ intersection of an orbit with a fixed plane ($\theta = 180°$ in the present case). This plot is called the *Poincaré return map*. It is surprising that the plot (r_n, r_{n+1}) defines a one-dimensional map as shown in figure 4.9. This reflects the fact that the original strange attractor has a structure of simple stretching and folding. It is noted that the structure of the ribbon-like attractor along the thickness is considered to be self-similar (see the footnote of p 86).

Now that a one-dimensional map with respect to (r_n, r_{n+1}) has been obtained from the return map above, the analysis of this one-dimensional map enables us to study the change of r_n as one makes a complete turn along the

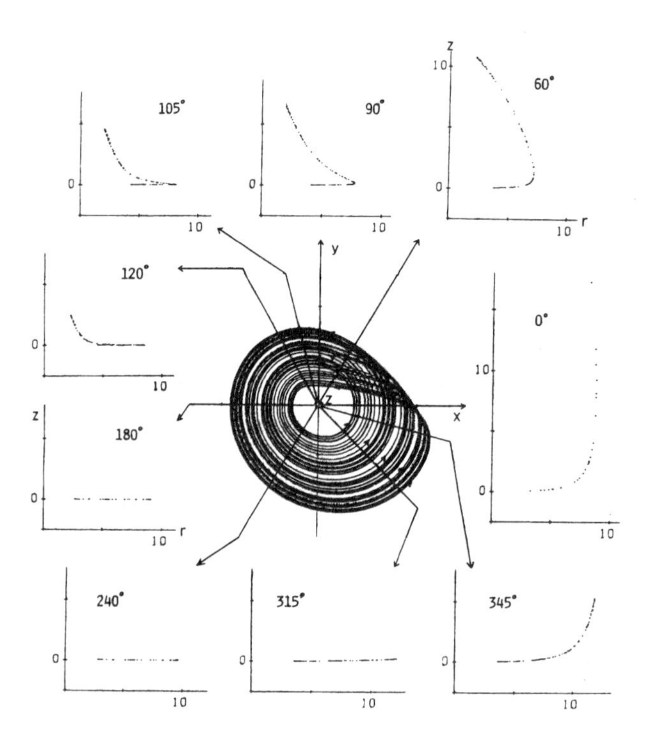

Figure 4.8. The Poincaré sections of the strange attractor of the Rössler model. Each Poincaré section is constructed with a half-plane whose edge is the z-axis. The Poincaré section is seen to be stretched and folded as one makes a complete turn around the z-axis.

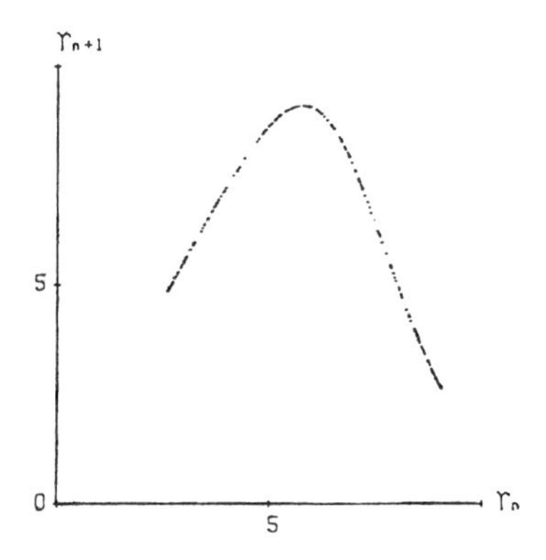

Figure 4.9. The return map (r_n, r_{n+1}) constructed from the Poincaré section with $\theta = 180^{\circ}$.

attractor. An essential point in this analysis is to approximate this map by some function. Let I be an interval over which the map is defined. One may introduce a single polynomial $f(x)$, for example, on the whole interval I, in which case, however, a polynomial of rather higher degree may be required and the approximation may be poor in some region of the interval. A better approach is to divide I into several subintervals and the map is approximated by a piecewise-polynomial function, which is called the *spline* function. Lyapunov numbers can be evaluated even for such a piecewise-polynomial function provided that it is differentiable at least once.

4.3 Lyapunov numbers and change of volume

In many cases, a strange attractor has a shape of a flat ribbon, like that of the Rössler model, reflecting the character of a dissipative dynamical system.

The rate of the change of a volume in the phase space, introduced in section 4.1, is given by

$$\frac{\dot{V}}{V} = \frac{\partial \dot{x}}{\partial x} + \frac{\partial \dot{y}}{\partial y} + \frac{\partial \dot{z}}{\partial z} = \frac{1}{5} + x - \mu \tag{4.22}$$

for the Rössler model. The long time average of $\frac{\dot{V}}{V}$ is negative since $\mu = 5.7$ here and x oscillates between positive and negative values (see figure 4.6) resulting in a vanishing contribution on average. This is the reflection of the fact that a region in the phase space is deformed into the attractor of a ribbon-like shape. Although the volume of the region decreases as a whole, it does not shrink uniformly but is stretched along one direction while it is squashed along the other direction. It will be explained in detail later that a region tends to be most stretched along the direction perpendicular to the orbit but still within the ribbon of the attractor in a similar manner as a one-dimensional map. If λ_1 denotes the Lyapunov number along the most stretched direction, the extension of the region along this direction expands as $e^{\lambda_1 t}$ in the long time average. Of course, λ_1 is positive. The region shrinks as $e^{\lambda_3 t}$ along the vertical direction to the ribbon, for which $\lambda_3 < 0$ (figure 4.10). The orbits are neither stretched nor squashed exponentially along the flow and hence $\lambda_2 = 0$ along this direction. The Rössler system being expressed in three variables x, y and z, there are three independent Lyapunov numbers which satisfy an inequality $\lambda_1 > \lambda_2 = 0 > \lambda_3$. The rate of the change of a volume as a whole is given by $\dot{V}/V = \lambda_1 + \lambda_2 + \lambda_3$, which turns out to be negative.

If a region in the phase space is stretched along a single direction while staying within a finite space, like the attractor of the Rössler model, there must be a mechanism to fold down the region. Thus there must be stretching and folding in the present case as in a one-dimensional map. Since the degree of stretching is measured by the Lyapunov number, it is reasonable to call a system with a positive Lyapunov number *chaotic*. The inequality $\lambda_1 > \lambda_2 = 0 > \lambda_3$ among the three Lyapunov numbers of the Rössler model (the set of Lyapunov

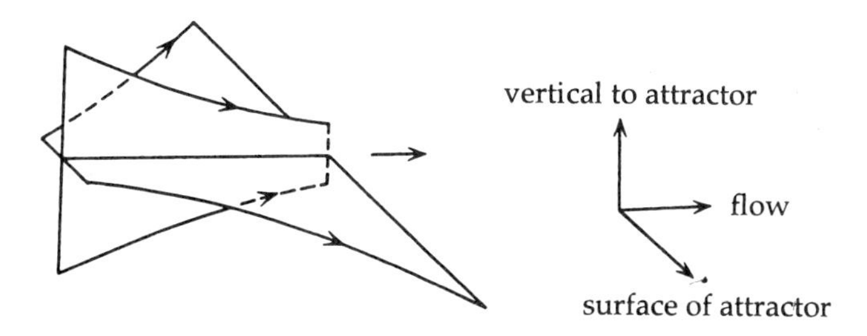

Figure 4.10. The separation between two orbits expands exponentially along one direction while it shrinks exponentially along the other direction. A flat attractor is formed in the presence of these flows.

numbers is called the Lyapunov spectrum) is a typical characteristic of a chaotic system with three variables, which also applies to the Lorenz model. Let us denote this set by $(+, 0, -)$. They satisfy $\sum_{i=1}^{3} \lambda_i < 0$ for a dissipative system while the Liouville theorem guarantees that $\frac{\dot{V}}{V} = \sum_{i=1}^{2N} \lambda_i = 0$ for a Hamilton system. Note also that the symmetry of a Hamilton system under time inversion $(t \to -t)$ implies $\lambda_i = -\lambda_{N-i+1}$.

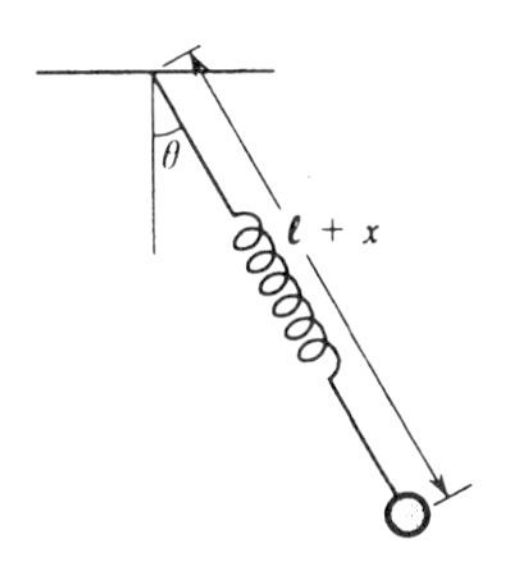

Figure 4.11. A pendulum whose string is a spring.

Let us consider a simple example of chaos in a Hamilton system. The equations of motion of a pendulum whose string is made of a spring (see figure 4.11) are given by

$$
\begin{aligned}
m\ddot{x} &= m(l+x)\dot{\theta}^2 + mg\cos\theta - kx \\
(l+x)\ddot{\theta} + 2\dot{x}\dot{\theta} &= -g\sin\theta.
\end{aligned}
\tag{4.23}
$$

Here l is the length of the spring in equilibrium, x is the displacement of the spring and θ is the angle between the spring and the vertical line. The system of equations (4.23) produces an irregular motion, an example of which is shown in figure 4.12.

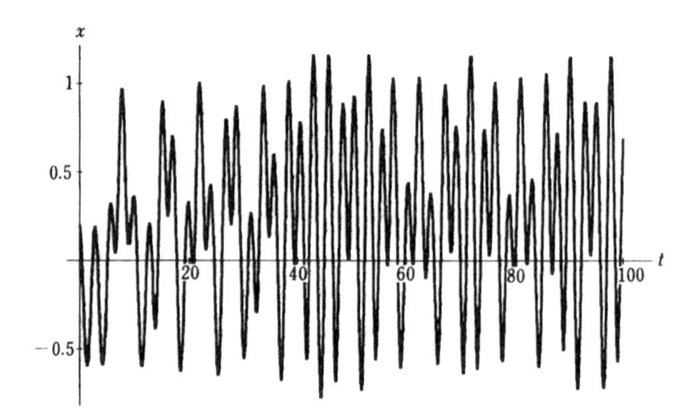

Figure 4.12. An irregular oscillation of a pendulum whose string is a spring. This shows x as a function of t.

A double pendulum, similar to the system considered here, will be analysed in appendix 4C.

Problem 5. Derive the equations of motion (4.23). Use the fact that the radial acceleration a_r and the angular acceleration a_θ perpendicular to a_r are given by

$$a_r = \ddot{r} - r\dot{\theta}^2 \qquad a_\theta = r\ddot{\theta} + 2\dot{r}\dot{\theta}. \tag{4.24}$$

Problem 6. Let the Hamiltonian of equations (4.23) be given by

$$H = \frac{y^2}{2m} + \frac{1}{2m}\frac{\alpha^2}{(l+x)^2} + mgl - mg(l+x)\cos\theta + \frac{1}{2}kx^2 \tag{4.25}$$

where $y = m\dot{x}$ and $\alpha = m(l+x)^2\dot{\theta}$. Verify the Hamilton equations of motion

$$\dot{x} = \frac{\partial H}{\partial y} \qquad \dot{y} = -\frac{\partial H}{\partial x} \qquad \dot{\theta} = \frac{\partial H}{\partial \alpha} \qquad \dot{\alpha} = -\frac{\partial H}{\partial \theta}. \tag{4.26}$$

Show also that $\dot{V}/V = 0$ in the phase space (x, y, θ, α).

For these Hamilton dynamical systems, it is required to study the properties of orbits starting from a domain of finite size in the phase space. In contrast, there exists a strange attractor in chaos of a dissipative system, such as the Lorenz model and the Rössler model, and all the orbits asymptotically approach the attractor at sufficiently later time. Therefore the study of the properties of the attractor reveals those of the chaotic orbits. There does not exist an attractor, in contrast, in a Hamilton system and the properties of chaos are not so simple as in a dissipative system.

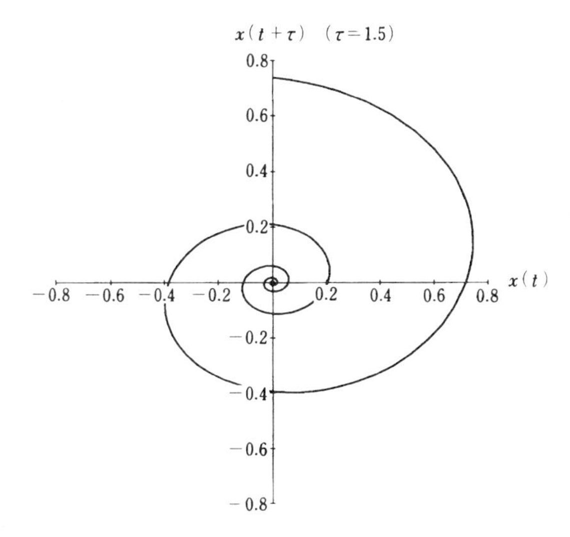

Figure 4.13. An orbit $(x(t), x(t+\tau))$ of a damped oscillator in the phase plane.

4.4 Construction of attractor

Let us consider how oscillating phenomena are measured in an actual experiment. The signal obtained from the system is of a single kind in many cases. In the cases of a simple harmonic oscillator or a damped oscillator, for example, the position and the velocity of the mass are not measured simultaneously but only the time dependence of the position is measured first, after which the velocity is obtained from the change of the position. Thus, once the continuous record $x(t)$ of the position is measured, the time derivative $\dot{x}(t)$ is determined, in principle, and hence the orbit $(x(t), \dot{x}(t))$ is obtained. In practice, however, differentiation of data tends to produce noise of high frequencies, which makes this operation *dirty*. It is, therefore, better to replace $\dot{x}(t)$ by a delayed signal $x(t+\tau)$. Since $x(t+\tau)$ contains a component linearly independent of $x(t)$, one can draw an orbit of a simple harmonic oscillator or a damped oscillator in the equivalent of the phase plane (see figure 4.13).

Problem 7. Let $x(t)$ be a solution to equation (4.12). Show that

$$x(t+\tau) = \mathrm{e}^{-r\tau}\left(\cos\tilde{\omega}\tau + \frac{r}{\tilde{\omega}}\sin\tilde{\omega}\tau\right)x(t) + \frac{\mathrm{e}^{-r\tau}}{\tilde{\omega}}\sin\tilde{\omega}\tau\,\dot{x}(t) \tag{4.27}$$

where $\tilde{\omega} = \sqrt{\omega_0^2 - r^2}$.

It is easy to draw this Lissajous pattern of the signals $x(t+\tau)$ and $x(t)$ with a help of digital memory.

Let us apply this method to measurements of general waves. Although a general system is usually nonlinear and the concept of linear independence is not applicable, we still employ our intuition obtained in the examples above.

To draw an orbit in an m-dimensional space from a single quantity $x(t)$, one takes an appropriate τ and constructs the orbit of

$$(x(t), x(t+\tau), x(t+2\tau), \ldots, x(t+(m-1)\tau))$$

as in figure 4.14. In this way, the orbit is embedded into an m-dimensional space. This m is called the *embedding dimension.* F Takens [12] has shown that characters of an attractor can be recovered by this embedding method.

The next question is how τ and m are determined in the embedding method. Although the size of τ may be arbitrary, in principle, the graph would be almost straight and any character would be lost if τ were too small to produce independent components, while the graph would be meaningless if τ were too large. Furthermore the noise would be amplified to yield unnecessary error, in the latter case, due to the orbital instability inherent in chaos. An empirical choice of τ is less than the average period by a factor. Figure 4.15 shows the attractor of the Rössler model reconstructed from the single variable $x(t)$ by this method. Also shown in figure 4.16 are the attractor reconstructed from a signal of the magnon chaos[2] and its Poincaré section. They reveal that the 'stretching and folding' mechanism is in action there.

As for m, it is necessary to take it to be at least 3 for an irregular oscillation. This may be determined by inspecting whether the correlation dimension defined in the next section takes a constant value.

The point attractor of the damped oscillation and the closed-curve attractor of the van der Pol equation are simple objects with dimension 0 and 1, respectively. The attractor of the Rössler model, on the other hand, has a two-dimensional structure of a thin ribbon type and the orbits form two-level crossings in a complicated manner. Accordingly this attractor is considered to have a fractal structure, explained in the next section. The structure perpendicular to the ribbon is not observed[3] in the Rössler model, whose attractor has strong two-dimensionality. Intuitively speaking, the attractor is thicker than a two-dimensional object since it has a thickness while it is by no means three dimensional and it should be rather called $(2+\alpha)$ dimensional $(0 < \alpha < 1)$. The attractor of frequently quoted difference equations

$$\begin{cases} x_{n+1} = y_n + 1 - ax_n^2 \\ y_{n+1} = bx_n \end{cases} \tag{4.28}$$

[2] A magnon (spin wave) excited in the ferromagnet $(CH_3NH_3)_2CuCl_4$ by the magnetic resonance induces a distinct oscillation with far lower frequency ($\sim$ 10 kHz) than the magnetic resonance frequency ($\sim$GHz), which turns out to be chaotic and is called the magnon chaos.

[3] Numerical computations of equation (4.20) yield the Lyapunov numbers of the Rössler model, namely $(0.072, 0.000, -5.388)$ for $\mu = 5.7$. The shrink rate along the thickness is of the order of $\simeq -\mu + \frac{1}{5}$. Since the average period is $\simeq 2\pi \simeq 6$, a structure extending over z =0–10 at $\theta = 60°$ in figure 4.8 shrinks to a microscopic size of $10\,e^{-5.4\times 6} \simeq 10^{-13}$ after executing a complete cycle.

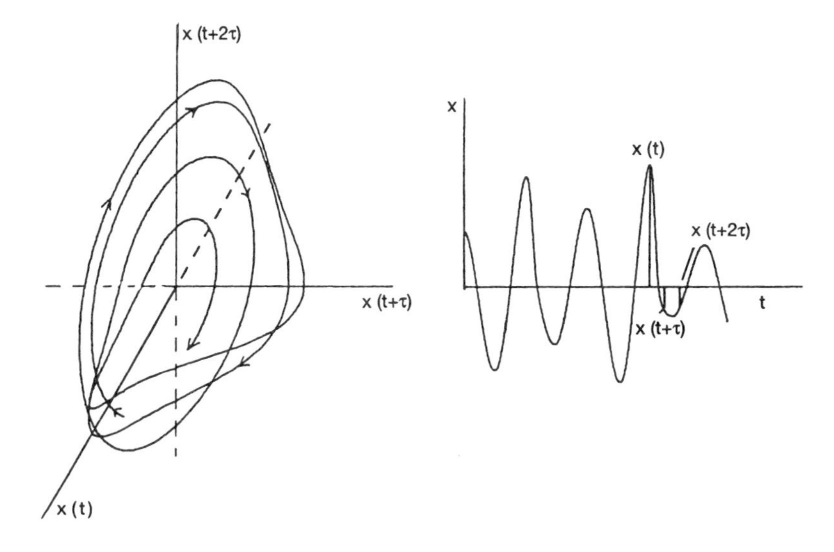

Figure 4.14. The attractor of the Rössler model reconstructed from the orbit $(x(t), x(t+\tau), x(t+2\tau))$.

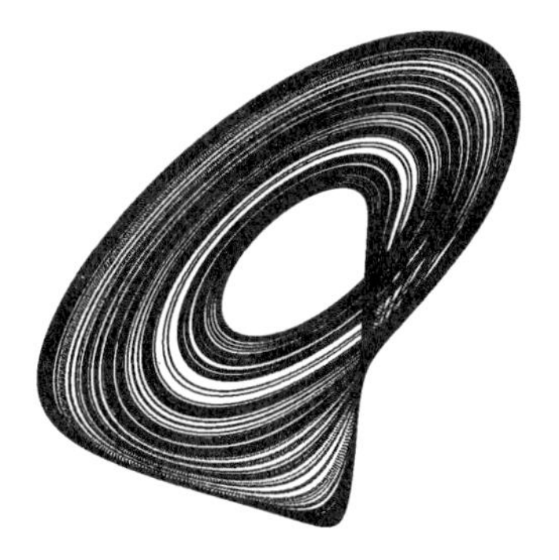

Figure 4.15. The attractor of the Rössler model reconstructed by the method employed in figure 4.13. (The projection on the $(x(t), x(t+\tau))$-plane.)

called the Hénon map, is shown in figure 4.17.

The dimensionality of such a figure with fractal structure will be considered in the next section.

The quickest way to obtain the nonintegral dimension of such a strange attractor by hand is the method of the correlation dimension introduced later.

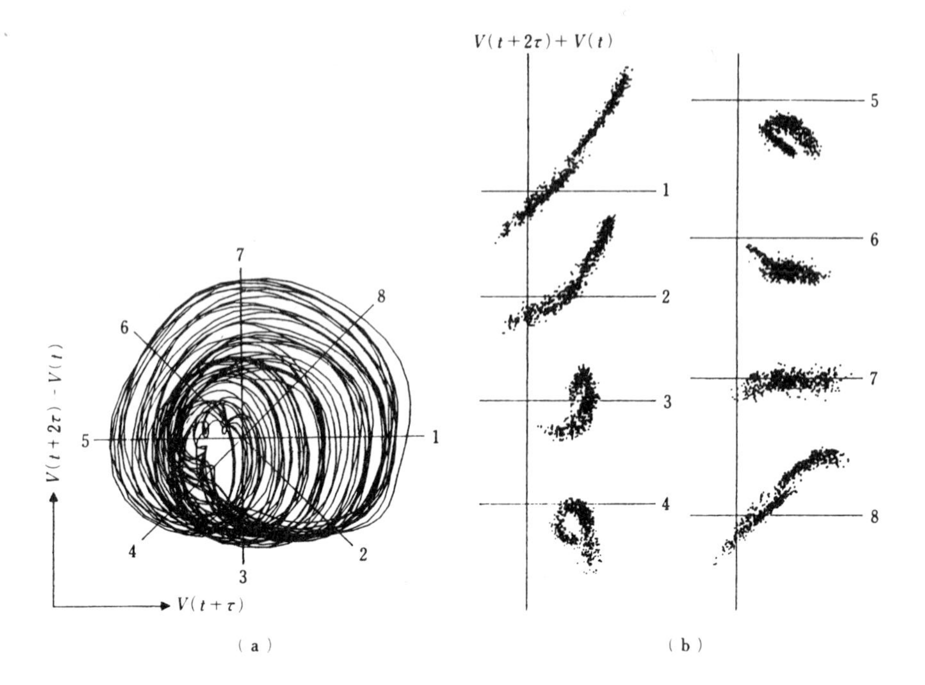

Figure 4.16. (a) The strange attractor obtained from the magnon-chaos signal and its Poincaré sections. The sections are perpendicular to the numbered lines. (b) Folding is observed in the sequence of the Poincaré sections 2 → 3 → 4 → 5 while stretching is seen in 6 → 7 → 8 → 1. (From Yamazaki H, Mino M, Nagashima H and Warden M 1987 *J. Phys. Soc. Japan* **56** 742.)

General methods to evaluate nonintegral dimension will be considered as follows.

4.5 Hausdorff dimension, generalized dimension and fractal

A typical definition of dimension that gives a nonintegral value is the Hausdorff dimension. Suppose a figure (a set) E is to be covered with closed sets[4] whose diameters[5] are less than ε (> 0). Let their diameters be $\varepsilon_1, \varepsilon_2, \varepsilon_3, \ldots$ and consider the infimum with respect to coverings,

$$\inf \sum_k \varepsilon_k^\alpha \quad (\alpha > 0). \tag{4.29}$$

[4] A closed set is, intuitively speaking, a set whose boundaries are part of the set. Among one-dimensional sets, closed intervals and Cantor sets are closed sets for example.

[5] The diameter of a set is defined as the distance between the two most distant points in the set. It is the diameter if the set is a sphere and the length of the diagonal line if it is a square.

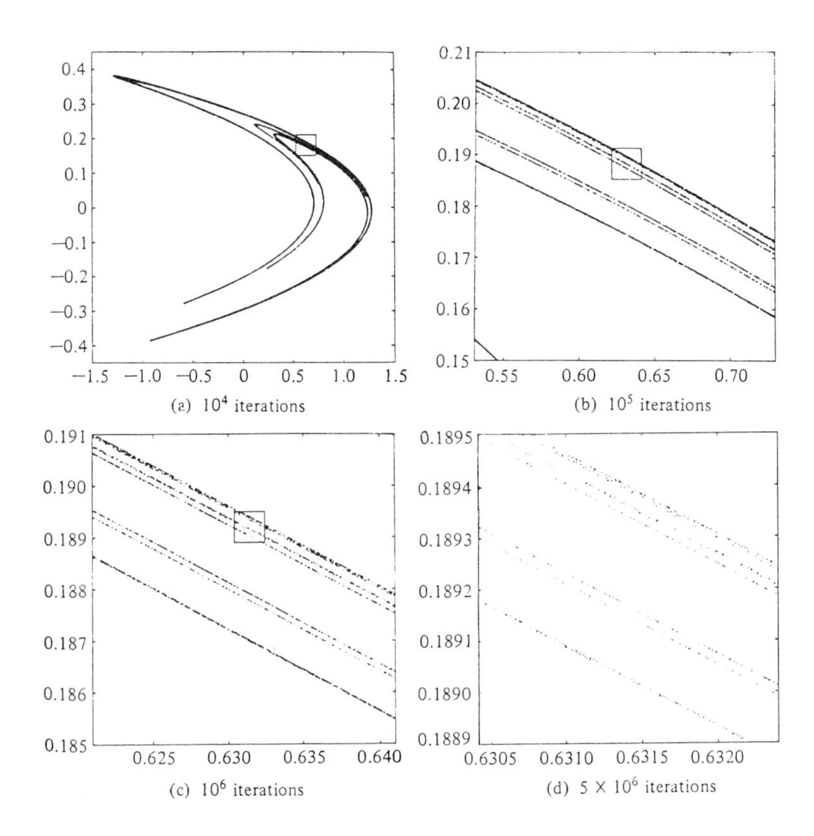

Figure 4.17. The attractor of the Hénon map. The square in the figure is magnified from (a) to (d). These figures reveal the self-similar structure of the attractor. (From Hénon M 1976 *Commun. Math. Phys.* **50** 69.)

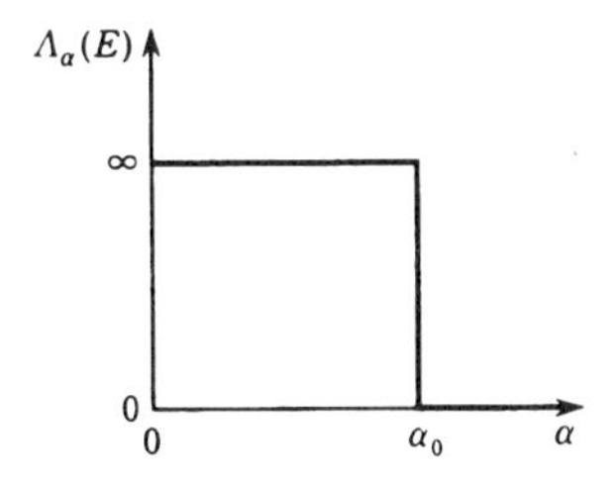

Figure 4.18. How to find the Hausdorff dimension of the set E.

In other words, we consider the most efficient covering which minimizes $\sum \varepsilon_k^\alpha$. Let the value of this summation be $\Lambda_\alpha(E)$ when $\varepsilon \to 0$. Then it can be shown that there exists a number α_0 such that

$$\Lambda_\alpha(E) = \begin{cases} \infty & (\alpha < \alpha_0) \\ 0 & (\alpha > \alpha_0) \end{cases} \tag{4.30}$$

(see figure 4.18). This number α_0 is called the *Hausdorff dimension* and denoted by D_0. The value of $\Lambda_\alpha(E)$ itself may take ∞, 0 or any number in between.

Let us consider simple examples.

(1) *An interval with the length l.* If this interval is covered with intervals with the common diameter ε (> 0), one needs at least l/ε intervals (strictly speaking this number is $[l/\varepsilon] + 1$, where [ ] is the Gauss symbol). Accordingly one has

$$\sum_{i=1}^{l/\varepsilon} \varepsilon^\alpha = l\varepsilon^{\alpha-1}.$$

When $\alpha > 1$, $\sum \varepsilon_k^\alpha$ is smaller if it is covered with smaller intervals, since $l\varepsilon'^{\alpha-1} < l\varepsilon^{\alpha-1}$ for $\varepsilon' < \varepsilon$ in this case. Since the number ε' may be arbitrarily small, one obtains $\inf \sum \varepsilon_k^\alpha = 0$ and hence $\lim_{\varepsilon\to 0} \inf \sum \varepsilon_k^\alpha = 0$. Let us consider the case $\alpha < 1$ next. For given ε, $\sum \varepsilon_k^\alpha$ takes the minimum value $l\varepsilon^{\alpha-1}$ if the interval is covered with spheres of the radius ε. Since $\inf \sum \varepsilon_k^\alpha = l\varepsilon^{\alpha-1}$, one obtains $\lim_{\varepsilon\to 0} \inf \sum \varepsilon_k^\alpha = \infty$. Thus the boundary is given by $\alpha = 1$ and hence the Hausdorff dimension of an interval of the length l is 1.

This example may also be understood in the following way. Since $\Lambda_\alpha = l$ if finite when $\alpha = 1$, one finds $D_0 = 1$. In general, when Λ_α takes a finite value for some $\alpha = \alpha_0$, the property (4.30) is satisfied for this α_0 and α_0 is identified as the Hausdorff dimension. Therefore, the parameter α_0 that gives a finite and fixed value $\Lambda_\alpha(E)$ may be employed as the Hausdorff dimension in practice.

Problem 8. Find the Haudorff dimension of a rectangle and a rectangular parallelepiped.

Thus, the Hausdorff dimension of such simple figures as an interval, rectangle and a rectangular parallelepiped agrees with an ordinary dimension.

(2) *A set of countably infinite points.* When $\alpha > 0$, these points may be covered with spheres with radii $\varepsilon, \varepsilon/2, \varepsilon/2^2, \ldots$, for which one has

$$\sum \varepsilon_k^\alpha = \frac{\varepsilon^\alpha}{1 - (1/2)^\alpha}.$$

This value is lowered if ε is replaced by ε' ($< \varepsilon$) and hence one has $\inf \sum \varepsilon_k^\alpha = 0$, which implies $D_0 = 0$. That is, the Hausdorff dimension of a countably infinite set is 0.

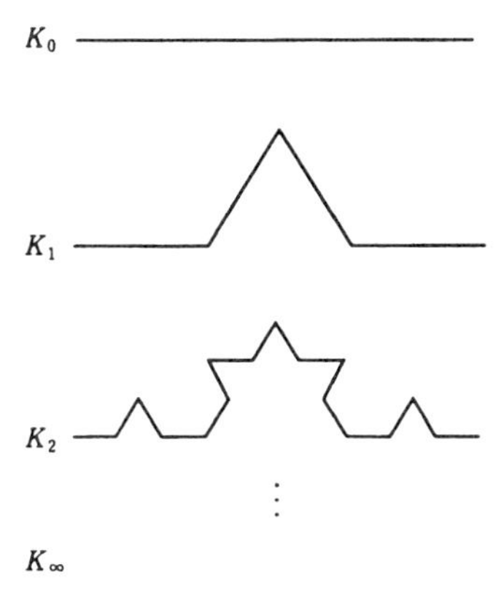

Figure 4.19. The procedure to generate the Koch curve K_∞.

(3) *The Koch curve.* The Koch curve is generated by the procedure described in figure 4.19. Let us take $\varepsilon = 1/3^n$ in the step K_n. One needs 4^n spheres with radius ε to cover the curve. Thus

$$\sum \varepsilon_k^\alpha = 4^n \left(\frac{1}{3^n}\right)^\alpha = e^{n(\log 4 - \alpha \log 3)}.$$

If one takes $\alpha_0 = \log 4 / \log 3$, one has $\sum \varepsilon_k^\alpha = 1$ and the above value becomes independent of n. Thus one obtains

$$\Lambda_{\alpha_0} = \lim_{\varepsilon \to 0} \inf \sum \varepsilon_k^\alpha = 1.$$

Therefore the Hausdorff dimension of the Koch curve is $D_0 = \alpha_0 = \log 4 / \log 3 = 1.2618\ldots$.

Figures, such as the Koch curve, that have nonintegral dimension are said to be *fractal* in general. Fractal figures tend to have a self-similar structure (magnification of a part of the figure contains the original figure) like the Koch curve and the Cantor set analysed below.

The capacity dimension is a simplified version of the Hausdorff dimension, where one sets $\varepsilon_1 = \varepsilon_2 = \ldots = \varepsilon$ in the definition of the Hausdorff dimension. If the minimum number of the spheres to cover the figure is denoted by $n(\varepsilon)$ and if one sets

$$\inf \sum \varepsilon_k^\alpha = n(\varepsilon)\varepsilon^\alpha \qquad \Lambda_{\alpha_0}(E) = C \ \ (= \text{finite})$$

one obtains

$$\alpha_0 = \lim_{\varepsilon \to 0} \frac{\log n(\varepsilon)}{\log \dfrac{1}{\varepsilon}}. \tag{4.31}$$

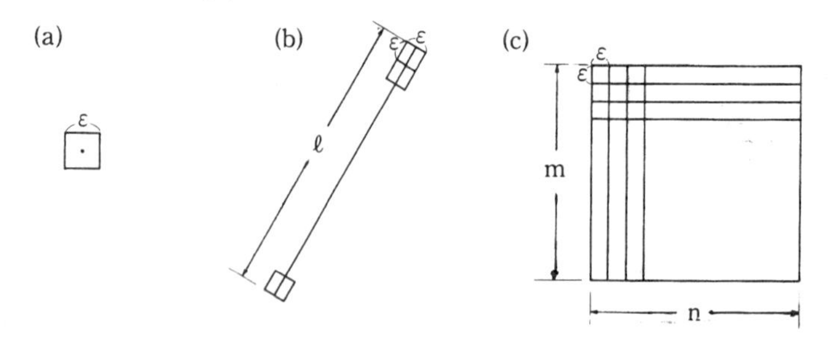

Figure 4.20. Capacity dimensions of a point, a segment and a rectangle.

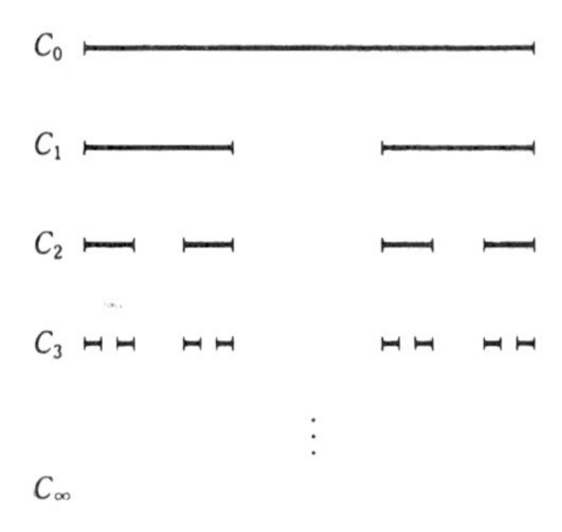

Figure 4.21. The procedure to generate the Cantor set C_∞. $C_0 = [0, 1], C_1 = C_0 - \left(\frac{1}{3}, \frac{2}{3}\right), C_2 = C_1 - \left(\frac{1}{9}, \frac{2}{9}\right) - \left(\frac{7}{9}, \frac{8}{9}\right), \ldots$.

This is called the *capacity dimension* and is denoted by D_{ca}.

The capacity dimension is understood intuitively and may be easily computed. Before proceeding to complex figures, let us look at the capacity dimension of an interval and a rectangle shown in figure 4.20 to facilitate our understanding. They have, of course, intergral capacity dimension, which agrees with ordinary dimension.

Let us consider the Cantor set next. Since $N(\varepsilon) = 2^n$ and $\varepsilon = 1/3^n$ in C_n of figure 4.21, one finds $D_{\text{ca}} = \log 2/\log 3 = 0.6309\ldots$.

These results on the Koch curve and the Cantor set are in agreement with our intuition. That is, the Koch curve should be somewhere between one dimension and two dimensions and the dimension of the Cantor set should be less than one, although the fractional part cannot be foreseen.

Problem 9. Let an interval cross another interval of the same length as shown in figure 4.22. Find the capacity dimension of the figure when this procedure is infinitely repeated.

The capacity dimension is applicable when a geometrical object is covered

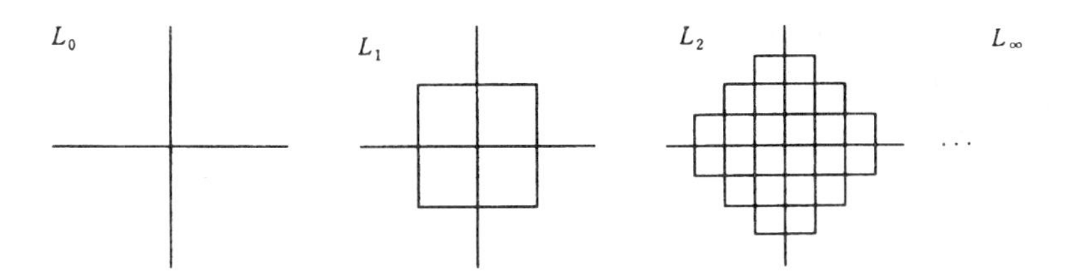

Figure 4.22. Construction of L_∞ from intervals. L_0: a cross made of two intervals of length 1, L_1: add to L_0 four intervals of length $1/2$ to form four crosses, L_2: add 16 intervals of length $1/4$ to obtain 16 crosses. L_∞ is obtained by repeating this procedure an infinite number of times.

with spheres (or cubes) of the same size in order and the number $n(\varepsilon)$ can be expressed as a function of ε as in the previous example.

Next appear the information dimension, correlation dimension and so forth. More general dimension is explained below before these dimensions are introduced independently.

Let us consider sets in the (d-dimensional) space under consideration. Let us divide this d-dimensional space into cubes with the edge ε and let $n(\varepsilon)$ be the total number of such cubes. Since we want to find the dimension of an attractor of chaos, let us find the probability p_i with which an orbit stays in the ith cube. A commonly employed method for this purpose is to sample the orbit $\boldsymbol{x}(t)$ with the time interval τ to construct the sample $(\boldsymbol{x}(0), \boldsymbol{x}(\tau), \boldsymbol{x}(2\tau), \ldots, \boldsymbol{x}((N-1)\tau))$ of the size N and count the number N_i $(i = 1, 2, \ldots, n(\varepsilon))$ of points in the ith cube to find the probability

$$p_i = \lim_{N\to\infty} \frac{N_i}{N} \tag{4.32}$$

with which the orbit is in the cube. The *generalized dimension* D_q is introduced with this p_i as

$$D_q = \lim_{\varepsilon\to 0} \frac{1}{q-1} \frac{\log\left(\displaystyle\sum_{i=1}^{n(\varepsilon)} p_i^q\right)}{\log \varepsilon} \qquad (-\infty < q < +\infty). \tag{4.33}$$

If one puts $q = 0$ in D_q, one finds

$$D_0 = \lim_{\varepsilon\to 0} \frac{\log n(\varepsilon)}{\log \dfrac{1}{\varepsilon}}$$

since $\sum_{i=1}^{n(\varepsilon)} p_i^0 = n(\varepsilon)$. This is the same as the capacity dimension defined by equation (4.31):

$$D_0 = D_{\mathrm{ca}}.$$

If one defines the case $q = 1$ by the limit $q \to 1$, one has

$$D_1 = \lim_{q\to 1} D_q = \lim_{\varepsilon\to 0} \frac{\sum_{i=1}^{n(\varepsilon)} p_i \log p_i}{\log \varepsilon}. \tag{4.34}$$

Problem 10. Use the L'Hôpital formula to derive equation (4.34).

The number D_1 is called the *information dimension* since the term $\sum p_i \log p_i$ in equation (4.34) has the same form as the information in the theory of information. Let us consider a simple example to understand the meaning of the information dimension. Consider the step C_1 in the construction of the Cantor set (figure 4.21). There are three boxes in this stage; suppose the probability of finding a point in boxes on both sides is $1/2$ and that in the central box is 0. Then one finds

$$\frac{\sum_{i=1}^{3} p_i \log p_i}{\log \varepsilon} = \frac{\frac{1}{2}\log\frac{1}{2} + 0\log 0 + \frac{1}{2}\log\frac{1}{2}}{\log\frac{1}{3}} = \frac{\log 2}{\log 3}. \tag{4.35}$$

This value is common for C_n with arbitrary n and the information dimension of the Cantor set is the same as its capacity dimension. Here $0 \log 0$ in equation (4.35) is understood in the sense of $\lim_{\delta\to 0} \delta \log \delta = 0$.

Problem 11. Show that the generalized dimension D_q of the Cantor set is independent of q and takes the value $\log 2/\log 3$.

The capacity dimension and the information dimension take the same value when the probability is independent of the position, as in the above example. If it depends on the position, in contrast, the information dimension is smaller than the capacity dimension in general. The data obtained by experiments depend on the position, and are processed accordingly as the latter case.

The choice $q = 2$ yields

$$D_2 = \lim_{\varepsilon\to 0} \frac{\log(\sum_{i=1}^{n(\varepsilon)} p_i^2)}{\log \varepsilon} \tag{4.36}$$

which is called the *correlation dimension* and is employed most frequently when the dimension is determined using a computer. An expression of the correlation dimension suitable for numerical computation will be given later.

It has been shown that there are many definitions of the dimension. Recently the generalized dimension has been often employed in computer simulations, in conjunction with the $f(\alpha)$ spectrum to be introduced in section 4.7. It can be shown (see appendix 4A) that $D_{q'} < D_q$ for $q < q'$, that is, D_q decreases monotonically with q.

Problem 12. Suppose $f(x)$ is convex upward and $\sum_{i=1}^{n} a_i = 1, a_i > 0$. Then

$$\sum_{i=1}^{n} a_i f(x_i) \leq f\left(\sum_{i=1}^{n} a_i x_i\right) \qquad \text{(Jensen's inequality)}. \tag{4.37}$$

Show that $D_0 \geq D_1 \geq D_2$ by choosing appropriate $f(x)$ above.

4.6 Evaluation of correlation dimension

One makes use of the correlation integral

$$C(\varepsilon) = \frac{1}{N^2} \sum_{i,j=1}^{N} H(\varepsilon - |\boldsymbol{x}_i - \boldsymbol{x}_j|) \tag{4.38}$$

to evaluate the correlation dimension D_2 of a strange attractor in practice. Here $H(\cdot)$ is the Heaviside function

$$H(x) = \begin{cases} 1 & (x \geq 0) \\ 0 & (x < 0) \end{cases} \tag{4.39}$$

and $|\boldsymbol{x}_i - \boldsymbol{x}_j|$ is the distance between $\boldsymbol{x}_i$ and $\boldsymbol{x}_j$ defined later. It is convenient to define the function

$$C_i(\varepsilon) = \frac{1}{N} \sum_{j=1}^{N} H(\varepsilon - |\boldsymbol{x}_i - \boldsymbol{x}_j|) \tag{4.40}$$

to specify the correlation integral and the correlation dimension. This is the probability of finding a point in the sphere of radius ε centred at $\boldsymbol{x}_i$. This means $p_i = C_i(\varepsilon)$. Therefore the summation $\sum p_i^2$ over the spheres may be replaced by the average over the position $\boldsymbol{x}_i$ as

$$\sum_{i=1}^{n(\varepsilon)} p_i^2 = \sum_{i=1}^{n(\varepsilon)} p_i C_i(\varepsilon) = \langle C_i(\varepsilon) \rangle = C(\varepsilon). \tag{4.41}$$

Thus one may write

$$D_2 = \lim_{\varepsilon \to 0} \frac{\log\left(\sum_{i=1}^{n(\varepsilon)} {p_i}^2\right)}{\log \varepsilon} = \lim_{\varepsilon \to 0} \frac{\log C(\varepsilon)}{\log \varepsilon}. \tag{4.42}$$

This means that $C(\varepsilon) \sim \varepsilon^{D_2}$.

The dimension of the strange attractor of the Rössler model is slightly larger than 2 if the above equation is employed. The numerical results and the example of the correlation dimension are shown next.

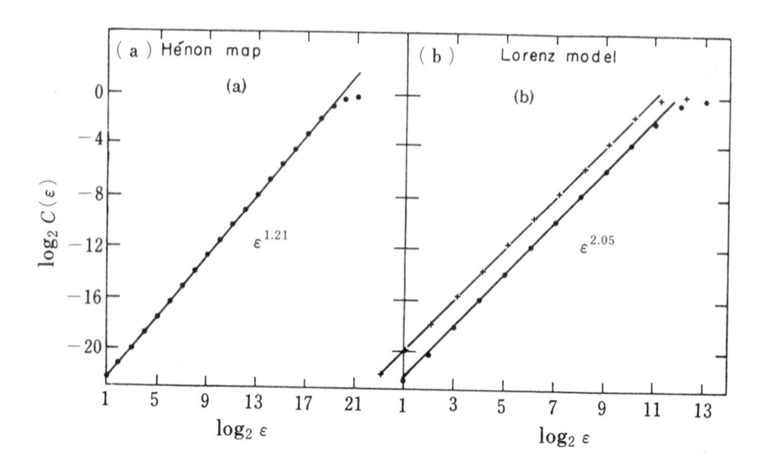

Figure 4.23. The correlation dimension obtained from the correlation integral. (a) The Hénon map and (b) the Lorenz model. The dimension obtained from the slope of the graph is close to 1 and 2, reflecting the shape of the respective attractor. (From Grassberger P and Procaccia I 1983 *Phys. Rev. Lett.* **31** 347.)

A point $\boldsymbol{x}_j = (x(t_j), x(t_j + \tau), \ldots, x(t_j + (m-1)\tau))$ in an m-dimensional space is constructed from the signal $x(t)$ to find the correlation dimension by the method of the correlation integral. Here $j = 0, 1, 2, 3, \ldots$. Most simply, the interval τ may be identical to the interval of $t_0, t_1, t_2, \ldots$. The correlation integral $C(\varepsilon)$ may be evaluated with these points using equation (4.38) and then D_2 is obtained from equation (4.42). If the embedding dimension m is less than the dimension D_2 of the attractor, the dimension cannot exceed m. The true dimension of the attractor is revealed when m is further increased.

Figures 4.23 and 4.24 show the dimension of the attractors of the Hénon map, the Lorenz model and the Rössler model. These values are close to 1, 2 and 2 respectively. Figure 4.25 shows the analysis of the correlation dimension of the strange attractor obtained from the magnon chaos mentioned previously.

Thus one obtains the dimension of the attractor from irregular experimental data. If this dimension is small, it suggests that the motion may be described by a small number of variables in practice, even when the original system has many degrees of freedom.

A few remarks on the correlation dimension analysis are in order. Although one may employ any distance $|\boldsymbol{x} - \boldsymbol{y}|$ between two vectors $\boldsymbol{x} = (x_1, x_2, \ldots, x_m)$ and $\boldsymbol{y} = (y_1, y_2, \ldots, y_m)$, it is most familiar to put $n = 2$ (the Euclid distance)

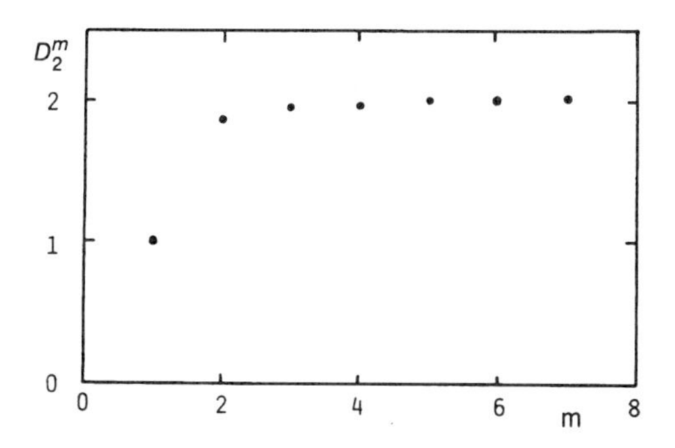

Figure 4.24. The correlation dimension D_2^m of the attractor of the Rössler model evaluated in the embedding space of the dimension m. The true correlation dimension D_2 reveals itself as m is increased. The value $D_2 \sim 2$ implies that the attractor has a ribbon shape similar to the Lorenz model.

in

$$|\boldsymbol{x} - \boldsymbol{y}| = \left(\sum_{i=1}^{m} |x_i - y_i|^n \right)^{1/n} \qquad (n = 1, 2, \ldots). \tag{4.43}$$

The distance with $n = 1$ is also often employed since the computation can be done faster. In fact, any definition of the distance gives the same result.

Next, one has to find the distribution of $|\boldsymbol{x}_i - \boldsymbol{y}_i|$ in each interval with the length $\log \varepsilon$ to find the relation between $\log C(\varepsilon)$ and $\log \varepsilon$. It is important to classify the distribution efficiently for computational purposes. For example, suppose each interval is assigned an integer k and increase the variable $A(k)$ in the computer by one each time $|\boldsymbol{x}_i - \boldsymbol{y}_i|$ hits this interval. In the end, the accumulated distribution of $|\boldsymbol{x}_i - \boldsymbol{y}_i|$, that is $C(\varepsilon)$, is obtained from $A(k)$, bypassing lengthy comparisons between numbers.

Next, consider the number N of the samples. If it is too small, the result will be unreliable while if it is too large, it takes too long for computations. In the case $D_2 = 2$, the number N may be several hundreds while N goes up as D_2 increases. If N is too large, however, it takes too long to compute $N(N-1)/2$ distances and classify them. An expedient is to draw M samples out of N and evaluate

$$\begin{aligned} C(\varepsilon) = \langle C_i(\varepsilon) \rangle &= \frac{1}{M} \sum_{i=1}^{M} C_i(\varepsilon) \\ &= \frac{1}{NM} \sum_{i=1}^{N} \sum_{j=1}^{M} H(x - |\boldsymbol{x}_i - \boldsymbol{x}_j|). \end{aligned} \tag{4.44}$$

The generalized dimension D_q may also be computed from $C_i(\varepsilon)$. One obtains,

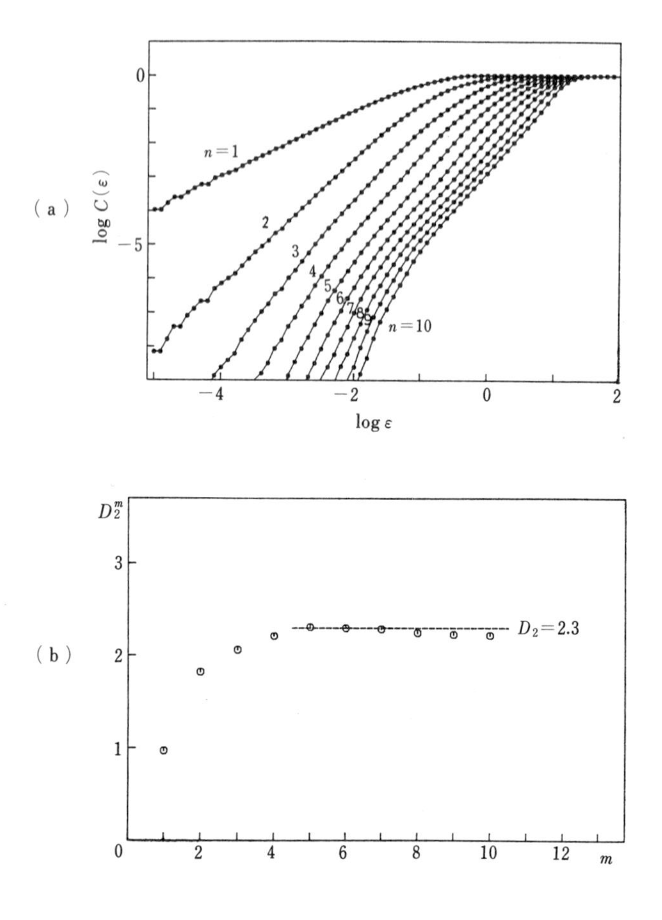

Figure 4.25. Experiments on the magnon chaos. (a) The relation between the correlation integral $C(\varepsilon)$ and ε. The correlation dimension is obtained from the portion $-4 < \log C(\varepsilon) < -2$. The part $\log C(\varepsilon) < -5$ is affected by noise. (b) This shows $D_2 \simeq 2.3$. (From Yamazaki H, Mino M, Nagashima H and Warden M 1987 *J. Phys. Soc. Japan* **56** 742.)

similarly to equation (4.41),

$$\sum_{i=1}^{n(\varepsilon)} p_i{}^q = \langle C_i^{q-1}(\varepsilon)\rangle \tag{4.45}$$

which, combined with equation (4.33), yields D_q. Concretely speaking, $(q-1)D_q$ is obtained from the slope of $\log\langle C_i^{q-1}(\varepsilon)\rangle$ as a function of $\log \varepsilon$, similarly to the correlation dimension.

Let us consider noise next. The signal obtained by experiments inevitably contains various kinds of noise. Since noise makes the orbit in the phase space fuzzy as much as the noise itself, one must estimate the dimension of the noise if ε is less than the noise level, while the noise may be neglected if ε is fairly significant compared to the noise level (see figure 4.26). Therefore it is necessary that the dimension may be estimated from $\log C(\varepsilon)$ and $\log \varepsilon$ in some interval of ε, where the noise level is low to a certain degree.

This method of the computation of the correlation dimension has been developed by P Grassberger and I Procaccia; its significance to experimenters is especially large since the number of data points is relatively small.

In the early days of experimental chaos, the statement 'chaos = randomness in a system of small degrees of freedom' was approved if (1) transition between regular and irregular oscillation is observed as a control parameter changes or (2) a one-dimensional map may be constructed by the Lorenz plot. In a realistic system, the change of the control parameter in (1) often requires a very fine tuning. Moreover, the experimental conditions must be well prepared beforehand. One has to make an enormous effort to tune parameters in an experimental system, which can be easily done on a computer. In other words, it is not easy to capture the route to chaos in experiment in general. The Lorenz plot in (2) is the planar plot of the points (A_i, A_{i+1}) in order $(i = 1, 2, 3, \ldots)$, that are constructed from the peak values $(A_1, A_2, \ldots, A_i, \ldots)$ of the signal obtained by experiments. This yields a one-dimensional map, as shown in the Rössler model, if it works well. However it is quite exceptional that a one-dimensional map is obtained by the Lorenz plot in general. That is, the attractor

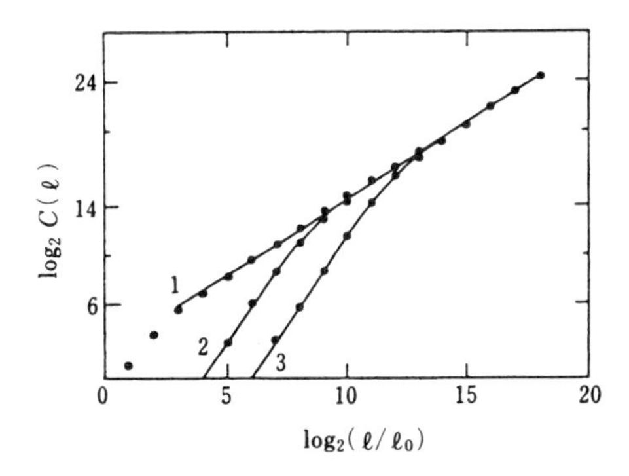

Figure 4.26. The correlation integral of the Hénon map affected by noise. The case 1 has no noise while the noise level increases in the order of 2 and 3. The dimension of the original Hénon map is revealed above the noise level. (From Beu-Mizrachi A and Procaccia I 1984 *Phys. Rev.* A **29** 975.)

has a two-dimensional structure as well as a simple 'folding and stretching' structure like that of the Rössler model. The Lorenz plot is effective only for such a simple chaos.

Compared to the methods mentioned above, the dimension of an attractor may be applicable to many systems. Given high-quality time-sequence data, it is possible to tell whether the signal has randomness with many degrees of freedom, like thermal noise, or has irregularity with few degrees of freedom. The Lyapunov number and the global spectrum $f(\alpha)$ introduced below may be evaluated if the quality of the data is pretty good. In fact, many data have been processed in this way and have contributed to research in chaos.

The relation between the dimension of an attractor and the Lyapunov number is considered next.

Let us consider an area of the attractor of a two-dimensional map (the Hénon map appearing previously, for example), surrounded by edges with lengths l_1 and l_2 as shown in figure 4.27. The lengths of the edges change to $l_1\,\mathrm{e}^{\lambda_1 t}$ and $l_2\,\mathrm{e}^{\lambda_2 t}$ respectively after time t has passed. Let D_1 and D_2 be the transverse and the longitudinal dimensions respectively. Since the number of points in this area is conserved, one obtains

$$l_1^{D_1} l_2^{D_2} = (l_1\,\mathrm{e}^{\lambda_1 t})^{D_1} (l_2\,\mathrm{e}^{\lambda_2 t})^{D_2}.$$

Therefore, one finds

$$D_1\lambda_1 + D_2\lambda_2 = 0 \qquad D_2 = -\frac{\lambda_1}{\lambda_2} D_1.$$

The total dimension D is

$$D = D_1 + D_2 = D_1 \left(1 - \frac{\lambda_1}{\lambda_2}\right).$$

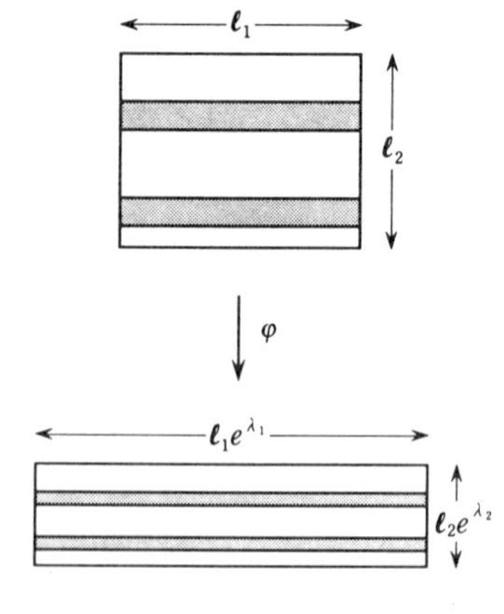

Figure 4.27. Modification of an area under a two-dimensional map φ.

D_1 is the dimension along which the attractor is linearly stretched while the attractor has a structure of the Cantor set along the D_2-direction. Therefore one may put $D_1 = 1$. That is,

$$D = 1 - \frac{\lambda_1}{\lambda_2} \tag{4.46}$$

where $\lambda_1 > 0$ and $\lambda_2 < 0$. This value is in good agreement with that computed directly in many two-dimensional maps [13]. J Kaplan and J Yorke have proposed the following general formula for the dimension of a strange attractor

$$D_{\mathrm{KY}} = j + \frac{\sum_{i=1}^{j} \lambda_i}{|\lambda_{j+1}|} \tag{4.47}$$

where $\sum_{i=1}^{j} \lambda_i \geq 0$ and $\sum_{i=1}^{j+1} \lambda_i < 1$ are assumed. This number D_{KY} is called the *Lyapunov dimension*.

4.7 Evaluation of Lyapunov number

It was shown for one-dimensional maps that the Lyapunov number measures the degree of stretching among the fundamental ingredients in chaos, i.e., stretching and folding. Since the function $f(x)$ is known exactly or approximately by making use of the spline function for a one-dimensional map, the Lyapunov number is easily evaluated from the derivative $f'(x)$. In contrast, the structure of the attractor cannot be given analytically for models, such as the Rössler model and the Lorenz model, that are expressed as a set of differential equations. Then the Lyapunov number, which is the measure of the motion on the attractor, must be evaluated observing how orbits are separated in time using a computer. It is important to realize then that at least one of the Lyapunov numbers is positive in a chaotic system and nearby orbits are separated exponentially in time and eventually cease to be nearby orbits. Therefore when nearby orbits are separated to a certain degree, one has to prepare another pair of nearby orbits to study the small deviation from the standard orbit as shown in figure 4.28. Suppose an initial difference vector $\boldsymbol{d}_0(0)$ develops to $\boldsymbol{d}_0(\tau)$ in a small time interval τ. Then the length of $\boldsymbol{d}_0(\tau)$ is reduced to that of $\boldsymbol{d}_0(0)$ keeping its direction fixed, resulting in a new vector $\boldsymbol{d}_1(0)$. That is,

$$\boldsymbol{d}_1(0) = |\boldsymbol{d}_0(0)| \cdot \frac{\boldsymbol{d}_0(\tau)}{|\boldsymbol{d}_0(\tau)|}.$$

Here $|\ |$ is the length of a vector. Vectors $\boldsymbol{d}_2, \boldsymbol{d}_3, \ldots$ are produced in order in this way. Since the nearby orbits are prepared in such a way that the difference vector points to the most developing direction, the expression

$$\lambda_1 = \lim_{N\to\infty} \frac{1}{N\tau} \sum_{i=0}^{N-1} \log \frac{|\boldsymbol{d}_i(\tau)|}{|\boldsymbol{d}_i(0)|} \tag{4.48}$$

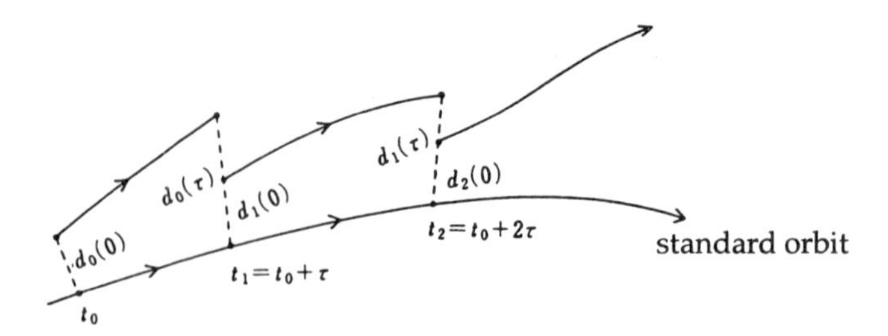

Figure 4.28. Evaluation of the largest Lyapunov number. To see the deviation of an orbit from the standard orbit under consideration, this orbit develops in time for a small interval τ. The distance is then reduced to $d_i(0)$ ($i = 1, 2, \ldots$) keeping the direction. The largest Lyapunov number is obtained from this procedure since the deviation of the nearby orbits develops toward the direction of the largest Lyapunov number.

gives the largest Lyapunov number.

The number of Lyapunov numbers is the same as the dimension of the phase space as noted before. Let us prepare the set of m mutually orthogonal initial vectors

$$(\boldsymbol{d}_0^{(1)}(0), \boldsymbol{d}_0^{(2)}(0), \ldots, \boldsymbol{d}_0^{(m)}(0))$$

and let them develop along m nearby orbits. Let $\boldsymbol{d}_0^{(1)}(0)$ be the vector that becomes largest among them and let $\boldsymbol{d}_1^{(1)}(0)$ be the vector $\boldsymbol{d}_0^{(1)}(\tau)$ reduced in amplitude. Then other vectors are no longer orthogonal to $\boldsymbol{d}_1^{(1)}(0)$ and must be made orthogonal by the Gram–Schmidt method as

$$\begin{aligned}
\boldsymbol{d}_{0\perp}^{(1)}(\tau) &= \boldsymbol{d}_0^{(1)}(\tau) \\
\boldsymbol{d}_{0\perp}^{(2)}(\tau) &= \boldsymbol{d}_0^{(2)}(\tau) - \frac{(\boldsymbol{d}_0^{(2)}(\tau), \boldsymbol{d}_0^{(1)}(\tau))}{|\boldsymbol{d}_{0\perp}^{(1)}(\tau)|^2}\boldsymbol{d}_{0\perp}^{(1)}(\tau) \\
\boldsymbol{d}_{0\perp}^{(3)}(\tau) &= \boldsymbol{d}_0^{(3)}(\tau) - \frac{(\boldsymbol{d}_0^{(3)}(\tau), \boldsymbol{d}_{0\perp}^{(1)}(\tau))}{|\boldsymbol{d}_{0\perp}^{(1)}(\tau)|^2}\boldsymbol{d}_{0\perp}^{(1)}(\tau) \\
&\quad - \frac{(\boldsymbol{d}_0^{(3)}(\tau), \boldsymbol{d}_{0\perp}^{(2)}(\tau))}{|\boldsymbol{d}_{0\perp}^{(2)}(\tau)|^2}\boldsymbol{d}_{0\perp}^{(2)}(\tau)
\end{aligned}$$

after which the amplitudes are adjusted. Here (,) denotes the inner product of vectors.

We simplify the notation as $\boldsymbol{d}_{0\perp}^{(2)}(\tau) = \boldsymbol{d}_\perp^{(2)}$ to write

$$\boldsymbol{d}_\perp^{(k)} = \boldsymbol{d}^{(k)} - a_{k,1}\boldsymbol{d}_\perp^{(1)} - a_{k,2}\boldsymbol{d}_\perp^{(2)} - a_{k,3}\boldsymbol{d}_\perp^{(3)} - \ldots - a_{k,k-1}\boldsymbol{d}_\perp^{(k-1)}. \tag{4.49}$$

Here

$$a_{k,i} = \frac{(\boldsymbol{d}^{(k)}, \boldsymbol{d}_\perp^{(i)})}{|\boldsymbol{d}_\perp^{(i)}|^2}.$$

Finally the basis vectors

$$d_1^{(i)}(0) = \frac{|d_0^{(i)}(0)|}{|d_{0\perp}^{(i)}(\tau)|} d_{0\perp}^{(i)}(\tau) \tag{4.50}$$

are constructed by normalizing the vectors $d_{0\perp}^{(1)}(\tau), d_{0\perp}^{(2)}(\tau), \ldots, d_{0\perp}^{(m)}(\tau)$. After repeating this procedure, one evaluates

$$\lambda_1 + \lambda_2 + \ldots + \lambda_m = \lim_{N\to\infty} \frac{1}{N\tau} \sum_{k=0}^{N-1} \log \frac{|d_k^{(1)}(\tau) \wedge d_k^{(2)}(\tau) \wedge \ldots \wedge d_k^{(m)}(\tau)|}{|d_k^{(1)}(0) \wedge d_k^{(2)}(0) \wedge \ldots \wedge d_k^{(m)}(0)|}$$

where $\wedge$ denotes the vector product and $|d^{(1)} \wedge d^{(2)} \wedge \ldots \wedge d^{(m)}|$ is the volume of the m-dimensional parallelpiped with the edges $d^{(1)}, d^{(2)}, \ldots, d^{(m)}$. By putting $m = 1, 2, 3, \ldots, n$ in the above equation, one obtains

$$\lambda_1, \lambda_1 + \lambda_2, \lambda_1 + \lambda_2 + \lambda_3, \ldots, \lambda_1 + \lambda_2 + \ldots + \lambda_n$$

in order, from which each Lyapunov number is determined.

The method mentioned so far has been developed by I Shimada and T Nagashima. Figure 4.29 shows the analysis of the Lyapunov numbers of the Lorenz model with this method. Table 4.1 shows how the Lyapunov numbers of the Rössler model converge as n is increased.

The methods developed so far are applicable when an equation generating chaos has been given. Suppose, in contrast, such an equation is not known beforehand as in the case of random signals obtained by experiments. Among several ideas to handle this problem, the method to evaluate the largest Lyapunov number, developed by S Sato, M Sano and Y Sawada (see figure 4.30), is explained here.

First the attractor is constructed with the method outlined in section 4.4 from the data obtained by experiments, after which this orbit is sampled with equal time intervals to obtain a set of points on the attractor. Find a pair of two points x_i and y_i whose distance is the shortest among the points in the set and follow the time evolution of these points and evaluate the ratio of the distances with time delay τ:

$$\Lambda_i(t, \tau) = \frac{|x_i(t+\tau) - y_i(t+\tau)|}{|x_i(t) - y_i(t)|}. \tag{4.51}$$

This is repeated for many pairs i, yielding

$$\langle \log \Lambda(t, \tau) \rangle = \frac{1}{N} \sum_{i=1}^{N} \log \Lambda_i(t, \tau). \tag{4.52}$$

The quantity corresponding to the Lyapunov number in the interval τ is defined as

$$\lambda(t, \tau) = \frac{1}{\tau} \langle \log \Lambda(t, \tau) \rangle. \tag{4.52'}$$

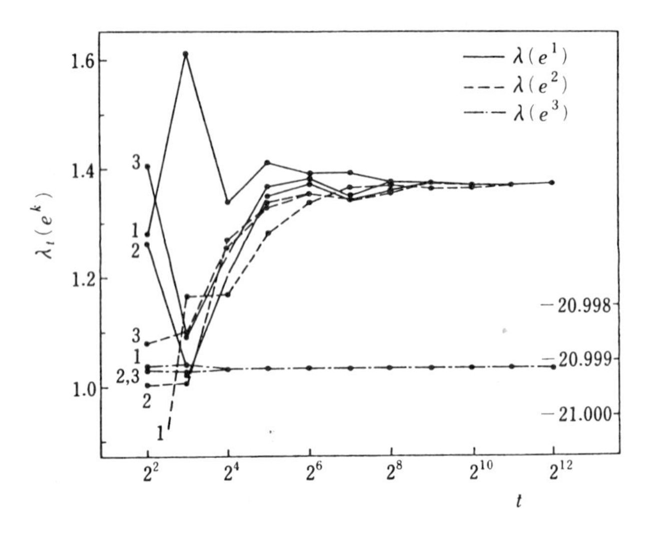

Figure 4.29. An example of Lyapunov numbers obtained from equations (4.49) and (4.50). The Lorenz model (4.21) is used with the parameters $\sigma = 16.0, b = 4.0$ and $r = 40.0$. The graph shows how $\lambda(e^1) = \lambda_1, \lambda(e^2) = \lambda_1 + \lambda_2$ and $\lambda(e^3) = \lambda_1 + \lambda_2 + \lambda_3$ converge with t. The result gives $\lambda_1 = 1.37, \lambda_2 = 0.00$ and $\lambda_3 = -22.37$. (From Shimada I and Nagashima T 1979 *Prog. Theor. Phys.* **61** 1605.)

Table 4.1. The number of steps in computation and convergence of the Lyapunov number in the Rössler model (4.20) with $\mu = 5.7$. The time step is 0.025.

n	λ_1	λ_2	λ_3
2	0.101 887	0.066 053	−4.701 911
4	0.077 659	0.105 560	−8.167 513
8	0.160 811	−0.006 103	−5.302 206
16	0.139 433	0.017 708	−6.122 024
32	0.125 435	0.004 908	−5.461 852
64	0.114 208	0.008 219	−5.600 146
128	0.094 953	0.002 615	−5.497 208
256	0.083 119	0.001 381	−5.437 145
512	0.090 574	−0.000 382	−5.396 670
1 024	0.081 649	−0.000 373	−5.399 836
2 048	0.076 505	−0.000 365	−5.393 630
4 096	0.073 376	−0.000 102	−5.388 497
8 192	0.072 709	−0.000 074	−5.388 979
16 384	0.071 962	−0.000 042	−5.387 793
32 768	0.071 989	−0.000 022	−5.387 947
65 536	0.071 931	0.000 005	−5.388 200
131 072	0.071 637	−0.000 001	−5.387 857

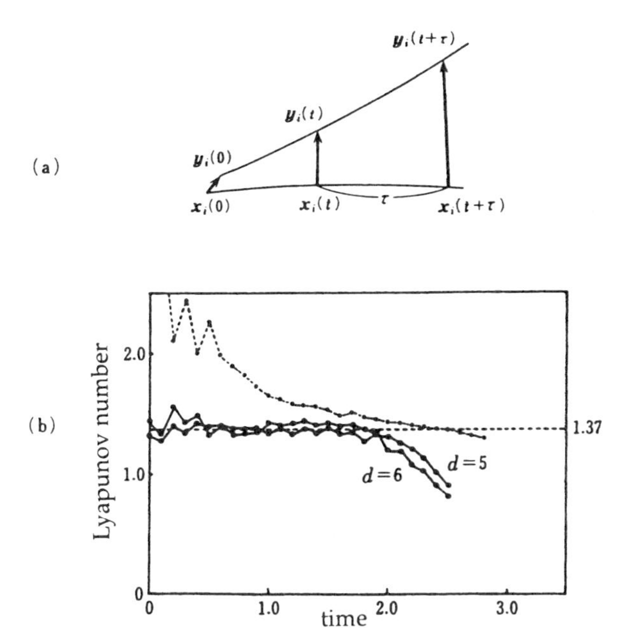

Figure 4.30. (a) A method to evaluate the largest Lyapunov number from experimental data. See equations (4.51) and (4.52). (b) The Lyapunov number of the Lorenz model has been obtained by this method. The numerical data are generated by the equation first and the method that is employed in the analysis of experimental data has been applied. The parameters in the mode are the same as those of figure 4.29. (From Sato S, Sano M and Sawada Y 1987 *Prog. Theor. Phys.* **77** 1.)

This serves as an estimate of the largest Lyapunov number (see figure 4.30).

The linearized form $\dot{\boldsymbol{\xi}} = T(\boldsymbol{x}(t))\boldsymbol{\xi}$ of an equation $\dot{\boldsymbol{x}} = \boldsymbol{F}(\boldsymbol{x})$ is considered in a more complicated method. Here T is a matrix constructed from $\boldsymbol{F}$ and its (ij)th matrix element is $T_{ij} = \partial F_i/\partial x_j$.

M Sano and Y Sawada [14] have given the method with which the most probable linearized equation may be guessed from experimental data.

Next, suppose a one-dimensional map is obtained from the Poincaré return map (x_i, x_{i+1}) constructed from positions x_i of a point by the Lorenz plot introduced in section 4.1, or by analysing the Poincaré section as done for the Rössler model. This map, obtained experimentally, is approximated by a single curve. Instead of employing a single polynomial function, which may introduce sizable discrepancy somewhere, the whole interval is divided into small intervals on each of which the map is approximated by a polynomial of a smaller degree, to fit with the curve everywhere on the interval. The neighbouring polynomials are matched at the boundary with the smoothness dictated by the degree of the

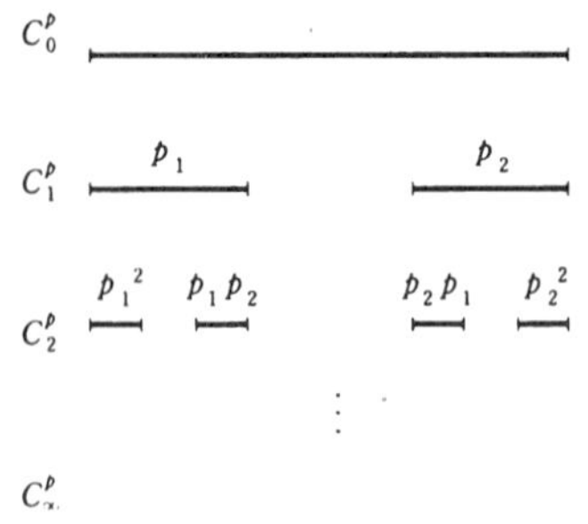

Figure 4.31. The construction of a Cantor set with position-dependent probability distribution. The choice $p_1 = 1/3$, $p_2 = 2/3$ is made in the following.

polynomials. By using the approximated $f(x)$, one finds from equation (2.18) the Lyapunov number

$$\lambda = \lim_{N\to\infty} \frac{1}{N} \sum_{i=0}^{N-1} \log |f'(x_i)|.$$

The use of $f(x)$ may be justified if the invariant measure is computed with a computer from this $f(x)$ and then the result is compared with experimental results.

It should be noted that a high quality (namely, good S/N (signal to noise ratio)) long term signal is required to estimate the dimension of the attractor, the Lyapunov numbers and the global spectrum to be introduced in the next section.

4.8 Global spectrum—the f(α) method

A method to quantify the whole picture of more complex fractal structures has been developed recently and has been applied not only to chaos but to many fields.

Let us start with a simple example. Consider a Cantor set whose probability distribution is position dependent as shown in figure 4.31. The generalized dimension D_q defined by equation (4.33) is obtained by considering $C_n{}^p$ as

$$D_q = \frac{1}{q-1} \frac{\log(p_1^q + p_2^q)^n}{\log l^n} = \frac{1}{q-1} \frac{\log(p_1^q + p_2^q)}{\log l}. \tag{4.53}$$

Figure 4.32 shows the graph of D_q as a function of q, where we have taken $p_1 = 1/3$, $p_2 = 2/3$ and $l = 1/3$. The figure shows that D_q is continuously distributed from $D_{-\infty}$ to D_∞ and there are contributions from small probability parts as $q \to -\infty$ and, in contrast, large probability parts are reflected upon

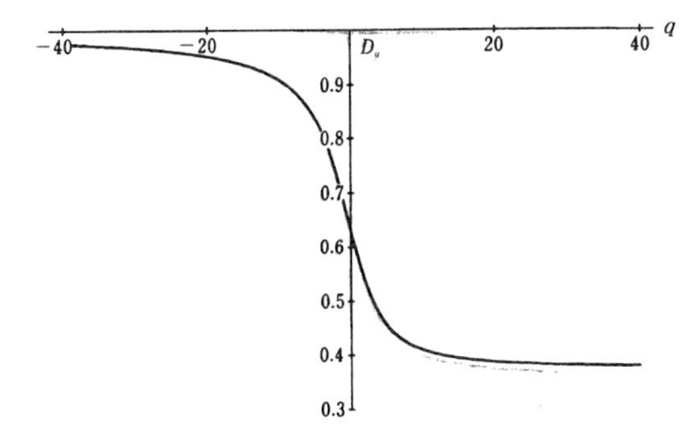

Figure 4.32. The generalized dimension D_q corresponding to figure 4.29.

the large q part. The q works as a parameter which filters parts with various probability.

The probability in the nth step in figure 4.31 is

$$p^{(n)} = p_1^m p_2^{n-m} \qquad (m = 0, 1, 2, \cdots, n)$$

from the right. Let us define here the length $l^{(n)}$ of the interval in the nth step and consider the quantity α defined by

$$p^{(n)} = (l^{(n)})^\alpha. \tag{4.54}$$

Then α has the meaning of the local dimension. This is because the probability is proportional to the length of the interval if $\alpha = 1$ while to the length squared if $\alpha = 2$. We are of course considering generalized nonintegral dimensions here. This parameter α is often called the *singularity* after T Halsey *et al* who proposed this analysis.

Next, let us express the distribution N of intervals with the singularity α as

$$N = (l^{(n)})^{-f(\alpha)}. \tag{4.55}$$

This is the capacity dimension of the elements with the dimension α and $f(\alpha)$ increases as N does. Let us find α and $f(\alpha)$ in this case. It follows from

$$l^{(n)} = l^n \ \ (l < 1) \qquad p^{(n)} = p_1^m p_2^{n-m} = (l^{(n)})^\alpha$$

that

$$\alpha = \frac{m \log p_1 + (n-m) \log p_2}{n \log l} = \frac{X \log p_1 + (1-X) \log p_2}{\log l} \tag{4.56}$$

where we have put $X = n/m$ $(0 \le X \le 1)$.

As for the number $N = \binom{n}{m}$, one finds from the Stirling formula

$$n! \sim \sqrt{2\pi n}\left(\frac{n}{e}\right)^n$$

that

$$\begin{aligned} N = \binom{n}{m} &= \frac{n!}{m!(n-m)!} \\ &\sim \sqrt{\frac{n}{2\pi m(n-m)}}\frac{n^n}{m^m(n-m)^{n-m}} \end{aligned} \tag{4.57}$$

which takes the maximum value at $m = 2/n$.

The asymptotic expression of $f(\alpha)$ for large n is

$$f(\alpha) = -\log N = \frac{X\log X + (1-X)\log(1-X)}{\log l}. \tag{4.58}$$

The parametric plot of $f(\alpha)$ with the parameter X is shown in figure 4.33 by employing these equations and putting $p_1 = \frac{1}{3}$, $p_2 = \frac{2}{3}$ and $l = \frac{1}{3}$. The singularity α in this case distributes between $\alpha_{\min}$ and $\alpha_{\max}$ defined by

$$\begin{aligned} \alpha_{\min} &= D_{\infty} = \frac{\log\frac{3}{2}}{\log 3} = 0.369\,07\ldots \\ \alpha_{\max} &= D_{-\infty} = \frac{\log 3}{\log 3} = 1. \end{aligned} \tag{4.59}$$

The $f(\alpha)$-curve for a Cantor set with a uniform probability distribution is a single point

$$\begin{aligned} \alpha &= \frac{\log p}{\log l} = \frac{\log 2}{\log 3} \\ f(\alpha) &= \frac{\log N}{\log l} = -\frac{\log 2^n}{\log(\frac{1}{3})^n} = \frac{\log 2}{\log 3}. \end{aligned} \tag{4.60}$$

It has been shown that the $f(\alpha)$-method is appropriate to reveal the global structure of a set with nonuniform distribution, yielding far more information compared to methods with a single dimension developed so far. There is inhomogeneity in the fractal structure obtained in model equations or measurements, in practice, and the distribution of dimension reveals the structure of the object as a whole. Therefore, this method has been applied not only to chaotic systems but to many fields where fractal structures appear. This method is called the *global spectrum method* or simply *$f(\alpha)$-method.*

In practical measurements, $f(\alpha)$ is obtained by reconstructing the attractor from the time sequence data and then the generalized dimension D_q is computed

D_0
0.6
0.5
0.4
0.3
0.2
0.1
D_∞
0.3 0.4 0.5 0.6 0.7 0.8 0.9 $D_{-\infty}$

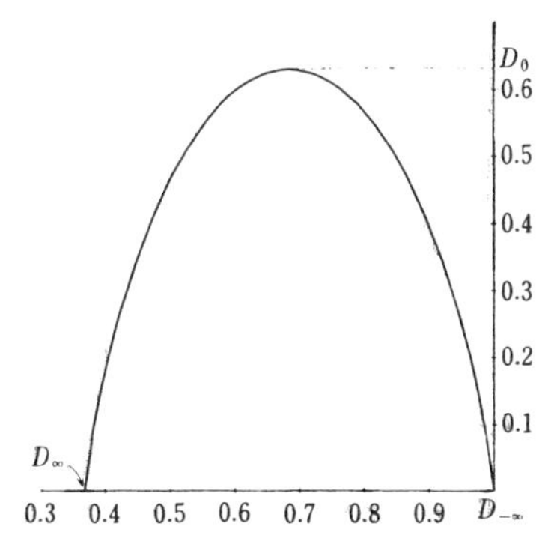

Figure 4.33. The $f(\alpha)$-spectrum for figure 4.31. One has the single point D_0 at $\alpha = 0$, in contrast, for an ordinary Cantor set with a uniform probability distribution.

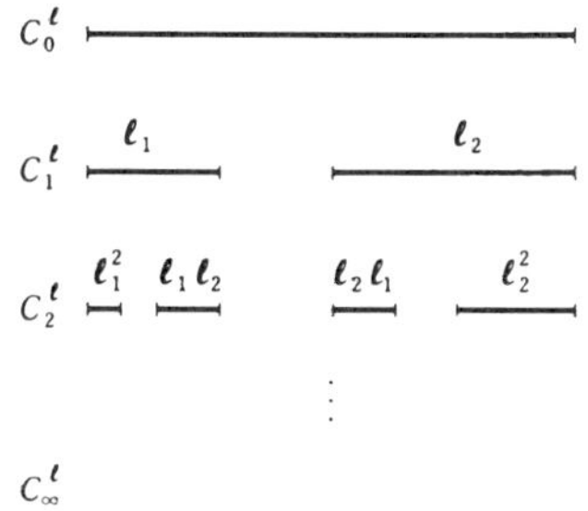

C_0^l
C_1^l
l_1
l_2
C_2^l
l_1^2
$l_1 l_2$
$l_2 l_1$
l_2^2
C_∞^l

Figure 4.34. Construction of the Cantor set C_∞^l with intervals of different lengths.

following equation (4.45) as explained before. Then $\alpha(q)$ and $f(\alpha(q))$ are evaluated from this D_q.

Let us next consider the Cantor set constructed from intervals with different lengths.

Problem 13. Show that

$$\alpha = \frac{\log p}{X \log l_1 + (1 - X) \log l_2}$$
$$f(\alpha) = \frac{X \log X + (1 - X) \log(1 - X)}{X \log l_1 + (1 - X) \log l_2} \tag{4.61}$$

in figure 4.34. Plot $f(\alpha)$ as a function of α for $p = 1/2, l_1 = 1/4$ and $l_2 = 1/2$.

One finds $\alpha_{\min} = \log 2/\log 4 = 0.5$ and $\alpha_{\max} = \log 2/\log 2 = 1$ in the above case.

Note that the generalized dimension D_q is not given by equation (4.33), that is applicable only to cases with equal division. The generalization is carried out as follows. Equation (4.33) may be rewritten as

$$\lim_{l\to 0} \frac{\sum p_i^q}{l^{(q-1)D_q}} = 1. \tag{4.62}$$

This equation may be generalized by introducing spheres with size $l_1, l_2, l_3, \ldots, l_n$ (<1) and considering the probability with which a point is in each sphere. Then evaluate τ that satisfies

$$\lim_{l\to 0} \left(\sum_{i=1}^{n} \frac{p_i^q}{l_i^\tau} \right) = 1$$

and find the dimension with the relation

$$\tau(q) = (q-1)\tilde{D}_q. \tag{4.63}$$

This generalized dimension $\tilde{D}_q$ agrees with D_q defined by equation (4.33) when $l_1 = l_2 = \ldots = l_n$ and it can be shown that $\tilde{D}_q$ is smaller than D_q otherwise. $\tilde{D}_q$ will be denoted simply as D_q henceforth.

Let us come back to the Cantor set. It follows from the self-similarity that the condition introduced above is written as

$$\lim_{n\to\infty} \left(\frac{p^q}{l_1^\tau} + \frac{p^q}{l_2^\tau} \right)^n = 1 \tag{4.64}$$

in the case of figure 4.31. That is, the condition to be satisfied is

$$\frac{p^q}{l_1^\tau} + \frac{p^q}{l_2^\tau} = 1.$$

For given parameters $p = 1/2, l_1 = 1/4$ and $l_2 = 1/2$, this equation is solved for τ by the Newton method to yield D_q for given q. Then one obtains a curve interpolating

$$D_{-\infty} = \frac{\log p}{\log l_1} = 0.5 \quad \text{and } D_\infty = \frac{\log p}{\log l_2} = 1.$$

Let us define the 'partition function' $\Gamma(q, l)$ by

$$\Gamma(q, l) = \sum_i p_i(l)^q. \tag{4.65}$$

Then one has

$$D_q = \lim_{l\to 0} \frac{1}{q-1} \frac{\log \Gamma(q, l)}{\log l}$$

whose behaviour for small l is

$$\Gamma(q,l) \sim l^{(q-1)D_q} = l^{\tau(q)}. \qquad (4.66)$$

Next consider the relation among D_q, α and $f(\alpha)$. The number of orbits $N(\alpha')\mathrm{d}\alpha'$ such that α takes a value in $[\alpha', \alpha' + \mathrm{d}\alpha']$ is given by

$$N(\alpha')\mathrm{d}\alpha' \sim \mathrm{d}\alpha' \rho(\alpha') l^{-f(\alpha')}$$

where $\rho(\alpha')$ is a suitable weight. Then $\Gamma(q,l)$ is written as

$$\Gamma(q,l) = \int l^{q\alpha'} l^{-f(\alpha')} \rho(\alpha')\,\mathrm{d}\alpha'.$$

The leading contribution to the above integral in the limit $l \to 0$ comes from such $\alpha' = \alpha(q)$ that minimizes $q\alpha' - f(\alpha')$. If $\Gamma(q,l)$ is evaluated with the saddle point method (see appendix 4B), one obtains

$$\Gamma(q,l) \simeq \sqrt{\frac{2\pi}{(\log l) f''(\alpha)}} \rho(\alpha) l^{q\alpha - f(\alpha)} \qquad (4.67)$$

where α is the solution to

$$\frac{\mathrm{d}}{\mathrm{d}\alpha'}(q\alpha' - f(\alpha')) = 0. \qquad (4.68)$$

Since this α is a function of q, it may be written as $\alpha = \alpha(q)$. Here it is assumed that such α is unique.

It follows from the assumption that

$$\frac{\mathrm{d}^2}{\mathrm{d}\alpha'^2}(q\alpha' - f(\alpha'))\Big|_{\alpha=\alpha(q)} > 0. \qquad (4.69)$$

From these equations, one obtains

$$f'(\alpha(q)) = q \quad \text{and} f''(\alpha(q)) < 0. \qquad (4.70)$$

Next, the dimension D_q is derived from

$$D_q = \lim_{l\to 0} \frac{1}{q-1} \frac{\log \Gamma(q,l)}{\log l}. \qquad (4.71)$$

If equation (4.67) is substituted into $\Gamma(q,l)$ and the identity $\lim_{l\to 0}[\log(\log l)/\log l] = 0$ is employed, one finds

$$D_q = \frac{1}{q-1}\{q\alpha(q) - f(\alpha(q))\} \qquad (4.72)$$

or

$$f(\alpha) = q\alpha - \tau(q). \tag{4.72'}$$

As for $\alpha(q)$, the relation

$$\alpha(q) = \frac{\mathrm{d}}{\mathrm{d}q}\{(q-1)D_q\} = \frac{\mathrm{d}}{\mathrm{d}q}\tau(q) \tag{4.73}$$

is derived.

Problem 14. Prove equation (4.73).

From the above observations, the $f(\alpha)$-curve is shown to possess the following properties:

(1) $f'(\alpha(0)) = 0$ and $f(\alpha(0)) = D_0$ for $q = 0$.
(2) $f(\alpha(1)) = \alpha(1) = D_1$ and $f'(\alpha(1)) = 1$ for $q = 1$. It follows from $f''(\alpha(1)) < 0$ that the curve $f = f(\alpha)$ is tangent to $f = \alpha$ at the point $(\alpha(1), f(\alpha(1))$ and $\alpha > f(\alpha)$ except at this point. This also implies

$$\frac{\mathrm{d}D_q}{\mathrm{d}q} = \frac{f(\alpha) - \alpha}{(q-1)^2} < 0 \quad (q \neq 1) \tag{4.74}$$

which shows D_q is a monotonically decreasing function of q.

Let us see how these relations are made use of by taking the Cantor set with different probability assignment defined in figure 4.31.

By substituting D_q of equation (4.53) into equation (4.73) and taking the derivative, one finds

$$\alpha(q) = \frac{p_1^q \log p_1 + p_2^q \log p_2}{(p_1^q + p_2^q)\log l}. \tag{4.75}$$

Substitution of this equation and D_q into equation (4.72) yields

$$f(\alpha(q)) = q\frac{p_1^q \log p_1 + p_2^q \log p_2}{(p_1^q + p_2^q)\log l} - \frac{\log(p_1^q + p_2^q)}{\log l}. \tag{4.76}$$

It can be shown that these results are in agreement with equations (4.56) and (4.58). This is clear from

$$f(\alpha(q)) = \frac{p_1^q \log p_1^q + p_2^q \log p_2^q - (p_1^q + p_2^q)\log(p_1^q + p_2^q)}{(p_1^q + p_2^q)\log l}$$

where we have put

$$X = \frac{p_1^q}{p_1^q + p_2^q}.$$

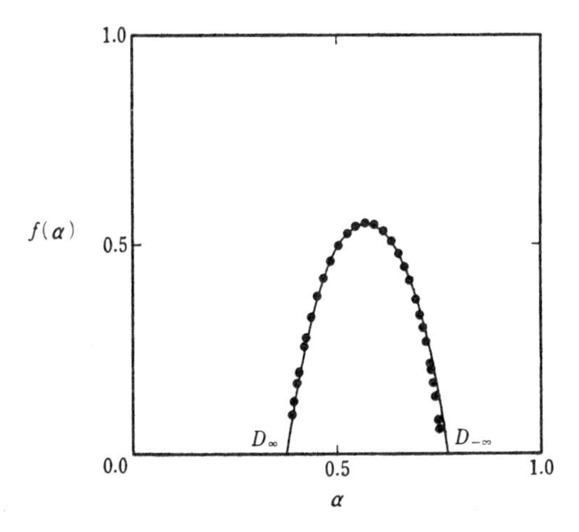

Figure 4.35. The $f(\alpha)$-spectrum of the logistic map $x_{n+1} = L_R(x_n) = Rx_n(1 - x_n)$ at $R = R_\infty = 3.569\,9456\ldots$ and the points obtained by magnon-chaos experiments. YIG is used as a magnetic material. (From Mitsudo S, Mino M and Yamazaki H 1992 *J. Magn. Magn. Mater.* **104–107** 1057.)

It follows from problem 14 that the saddle point method leading to equation (4.72) and the exact solution with the Stirling formula give the same conclusion.

Problem 15. Apply the saddle point method in the definition of the gamma function

$$\Gamma(n+1) = \int_0^\infty \mathrm{e}^{-x} x^n \, \mathrm{d}x$$

to prove the Stirling formula

$$n! \sim \sqrt{2n\pi} \left(\frac{n}{\mathrm{e}}\right)^n .$$

Figure 4.35 shows the $f(\alpha)$-curve of the set at an accumulation point of a pitchfork bifurcation point of the logistic map and the points obtained from magnon-chaos experiments as a practical application of the $f(\alpha)$-method. α in this figure is shown to lie between

$$\alpha_{\min} = D_\infty = \frac{\log 2}{\log \alpha_{\mathrm{PD}}^2} = 0.3775\ldots$$

and (4.77)

$$\alpha_{\max} = D_{-\infty} = \frac{\log 2}{\log \alpha_{\mathrm{PD}}} = 0.7555\ldots$$

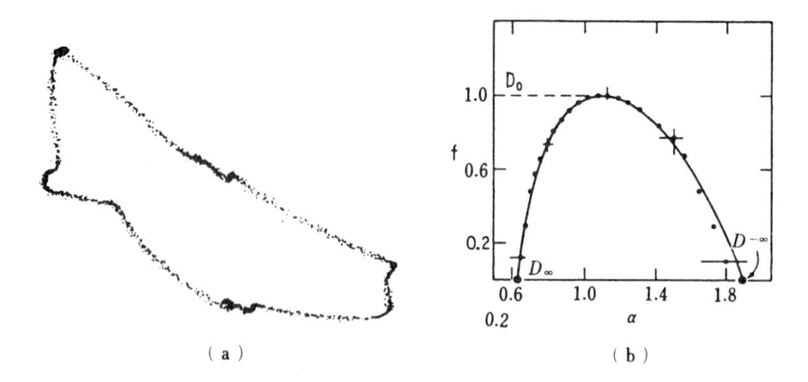

Figure 4.36. (a) The two-dimensional attractor and (b) the $f(\alpha)$-characteristics, obtained from Benard convection experiments. Dots are obtained from experiments while the solid curve is a theoretical prediction (From Jensen M H, Kadanoff L P, Procaccia I and Stavans J 1985 *Phys. Rev. Lett.* **55** 2798.)

which follows from the period-doubling constant

$$\alpha_{\mathrm{PD}} = 2.502\,907\,875\ldots$$

introduced in chapter 3.

Problem 16. It follows from the explanation of the period-doubling phenomena in Chapter 3 that the set of accumulation points is represented by the Cantor set with different intervals, shown in figure 4.34, with the choice $l_1 = 1/\alpha_{\mathrm{PD}}$ and $l_2 = 1/\alpha_{\mathrm{PD}}^2$ (albeit the order of the elements is different). Find the $f(\alpha)$ curve in this case and compare the result with figure 4.35.

Before we close this chapter, an attempt initiated by M H Jensen *et al* on the observation of $f(\alpha)$ in experiments is explained. The system under study is the Benard convection[6] in mercury under external electromagnetic force. The frequency of the external force is taken to be the golden ratio $(=(\sqrt{5}-1)/2)$ of that of the original Benard convection. The attractor in this case is shown in figure 4.36(a). Figure 4.36(b) shows the points of $f(\alpha)$ obtained from this attractor. The attractor is divided into intervals with the length l and $\Gamma(q,l)$ of equation (4.65) is evaluated as the average over the orbits as

$$\Gamma(q,l) = \langle p_i(l)^{q-1} \rangle.$$

[6] Convection takes place when a liquid is heated from below. There appears a spatial pattern determined by the surface or the shape of the container if it is shallow which is called the Benard convection. There appears a stable roll-shaped pattern if the upper surface is a solid boundary. This roll is unstable and makes a transition to chaos as the temperature difference becomes large.

This average may be replaced by the inverse of the recurrent time m_i for an orbit to come back to the interval which the point started from. Then $\tau(q)$ may be obtained from this and equation (4.66), namely,

$$\Gamma(q, l) = \langle m_i^{1-q} \rangle \sim l^{\tau(q)}.$$

Once this $\tau(q)$ is obtained as a function of q, one finds the relation between α and $f(\alpha)$ from

$$\alpha(q) = \frac{\mathrm{d}}{\mathrm{d}q}\tau(q) \quad \text{and } f(\alpha) = q\alpha(q) - \tau(q).$$

The experimental results are in good agreement with the analysis based on the circle map (the real line).

Appendices

1A Periodic solutions of the logistic map

Periodic solutions of the map $x_{n+1} = 4x_n(1 - x_n)$ are obtained as follows.
Put $x_n = \sin^2\theta_n$ to find

$$\begin{aligned}\sin^2\theta_{n+1} &= 4\sin^2\theta_n(1 - \sin^2\theta_n)\\ &= 4\sin^2\theta_n\cos^2\theta_n = \sin^2 2\theta_n.\end{aligned}$$

The point x_{n+p} obtained after p iterations of x_n is given by

$$\sin^2\theta_{n+p} = \sin^2 2^p\theta_n$$

from which one finds

$$\theta_{n+p} = \pm 2^p\theta_n + m\pi.$$

If the orbit has period p, x_n satisfies the condition

$$x_{n+p} = \sin^2\theta_{n+p} = x_n = \sin^2\theta_n$$

namely,

$$\begin{aligned}&\theta_{n+p} = \pm\theta_n + m'\pi \quad \text{i.e.,} \quad \pm\theta_n + m'\pi = \pm 2^p\theta_n + m\pi\\ &\text{i.e.,} \quad (2^p \pm 1)\theta_n = (m - m')\pi = l\pi \quad \text{i.e.,} \quad \theta_n = \frac{l\pi}{2^p \pm 1}.\end{aligned}$$

Let us take $p = 2$ and $l = 1$, for example. Then

$$\theta_n = \frac{\pi}{2^2 \pm 1} = \begin{cases} \dfrac{\pi}{3}; & x_1 = \sin^2\dfrac{\pi}{3} = 0.75 \quad \text{(period 1)}\\[2ex] \dfrac{\pi}{5}; & x_1 = \sin^2\dfrac{\pi}{5} = 0.345\,491\,5028\\[2ex] & x_2 = 0.904\,508\,4972 \quad \text{(period 2).}\end{cases}$$

The period 3 solution is found by putting $p = 3$ and $l = 1$ as

$$\theta_n = \frac{\pi}{2^3 \pm 1} = \begin{cases} \frac{\pi}{7}; & x_1 = \sin^2\frac{\pi}{7} \quad x_2 = \sin^2\frac{2\pi}{7} \quad x_3 = \sin^2\frac{4\pi}{7} \\ \frac{\pi}{9}; & x_1 = \sin^2\frac{\pi}{9} \quad x_2 = \sin^2\frac{2\pi}{9} \quad x_3 = \sin^2\frac{4\pi}{9} \end{cases}$$

where

$$x_1 = \sin^2\frac{\pi}{7} = 0.188\,255\,0991 \text{ and } x_1 = \sin^2\frac{\pi}{9} = 0.116\,977\,784$$

respectively.

2A Möbius function and inversion formula

Let functions f and g be defined on the set of natural numbers and suppose they satisfy the relation

$$g(n) = \sum_{d|n} f(d)$$

for any $n \geq 1$. Here the summation is taken over all d that divide n. The inverse of this relation is obtained by making use of the Möbius function as

$$f(n) = \sum_{d|n} \mu\left(\frac{n}{d}\right) g(d)$$

which is called the *Möbius inversion formula.* The Möbius function is defined on the set of natural numbers and takes the value $\mu(1) = 1$, $\mu(n) = (-1)^k$ if n is the product of k (≥ 1) mutually distinct prime numbers $p_1 p_2 \cdots p_k$ and, finally, $\mu(n) = 0$ if n is divisible by a square of an integer. For example, $\mu(3) = -1, \mu(6) = 1, \mu(9) = 0$ and $\mu(12) = 0$. The proof of the inversion formula may be found in a book on elementary number theory (see [15] for example).

In the case of the tent map $T(x)$, there are 2^n solutions to the equation $T^n(x) = x$. Therefore if one denotes the total number of period d points by $A(d)$, one obtains

$$\sum_{d|n} A(d) = 2^n.$$

Now one takes $g(n) = 2^n$ and $f(d) = A(d)$ in the inversion formula to find

$$A(n) = \sum_{d|n} \mu\left(\frac{n}{d}\right) 2^d.$$

2B Countable set and uncountable set

An uncountable set is an infinite set that is not countable. What is a countable set, then? It is necessary to introduce the concept of the 'potency' of a set to answer this question. If the set under consideration is a finite set, its *potency* is the total number of the elements of the set. Accordingly, two sets have the same potency if they have the same number of elements. When the sets are infinite, they are considered to be of the same potency if there is a one-to-one correspondence between the elements of the two sets 'under a certain rule'. For example there is a one-to-one correspondence between the set of natural numbers $\boldsymbol{N} = \{1, 2, 3, \ldots, n, \ldots\}$ and the set of positive even numbers $\{2, 4, 6, \ldots, 2n, \ldots\}$ under the correspondence 1 and 2, 2 and 4, 3 and 6, ..., n and $2n$, ... Therefore these two sets have the same potency. (One might think the potency of $\boldsymbol{N}$ is twice as much as that of the even numbers. Thus the situation for infinite sets is quite different from that of finite sets.) It can be shown, in the same way, that the set $\boldsymbol{N}$ and the set of squared integers $\{1, 4, 9, \ldots, n^2, \ldots\}$ have the same potency by introducing correspondence between n and n^2. It was G Galilei (1564–1642) who considered such a correpondence first. Later G Cantor (1845–1918) elaborated this idea and constructed the theory of infinite sets.

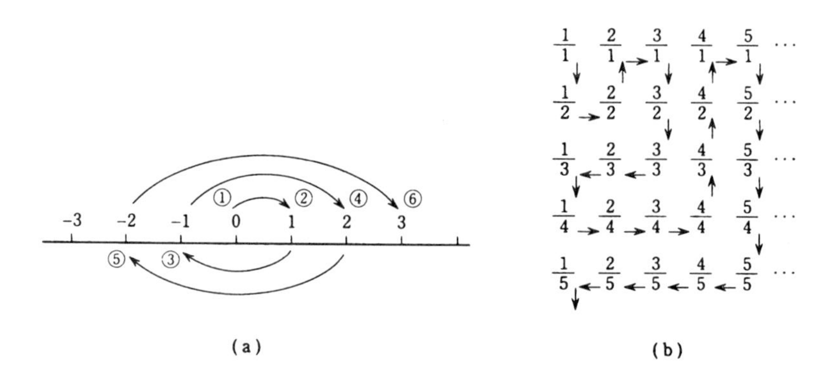

Figure 2B.1. (a) Numbering integers. (b) Numbering of positive rational numbers. (Numbering of irreducible fractions may be similarly done.)

Examples of the set with the same potency as $\boldsymbol{N}$, besides the two introduced above, include the set of integers, the set of rational numbers and the set of algebraic numbers.[1]

Existence of a one-to-one correspondence between a set and the set of

[1] The solutions of the equation $a_0x^n + a_1x^{n-1} + \ldots + a_{n-1}x + a_n = 0$ $(n \geq 1)$ with integral coefficients are called algebraic numbers. The set of algebraic numbers includes integers, rational numbers and roots of rational numbers.

natural numbers $\boldsymbol{N}$ implies the elements of the former set may be counted (numbered) as $1, 2, 3, \ldots$. Accordingly such a set is called a *countable set.* A countable set has the lowest potency among infinite sets. The potency of an infinite set which comes next to that of a countable set is considered to be the potency of the set of real numbers $\boldsymbol{R}$. A set with this potency is called a *continuum.*[2] There are also some facts against our common sense. The potency of the real line $\boldsymbol{R}$ is equal to that of an interval [0, 1], a square ($[0, 1] \times [0, 1] = [0, 1]^2$) and a cube $[0, 1]^3$ and so on.

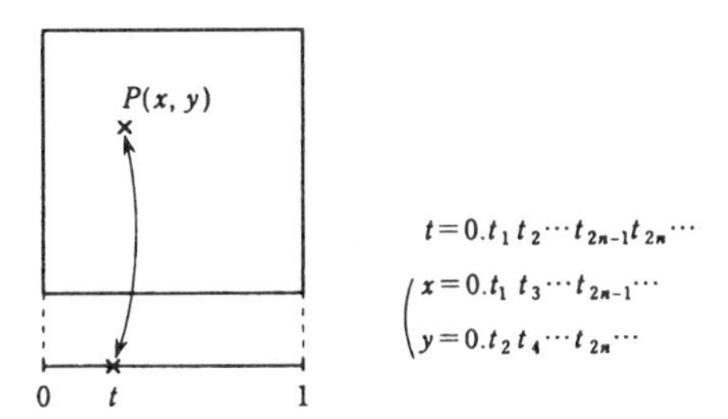

Figure 2B.2. An interval [0, 1] has a one-to-one correspondence with a square $[0, 1] \times [0, 1]$.

The scrambled set S that appears in the Li–Yorke theorem is an uncountable subset of $\boldsymbol{R}$ and hence S is a continuum. Another example of a continuum is the *Cantor set.* The Cantor set is constructed by equally dividing an interval [0, 1] into three and removing the central part $\left(\frac{1}{3}, \frac{2}{3}\right)$ and then equally dividing each of the residual two intervals $\left[0, \frac{1}{3}\right]$ and $\left[\frac{2}{3}, 1\right]$ into three and removing the open intervals $\left(\frac{1}{9}, \frac{2}{9}\right)$ and $\left(\frac{7}{9}, \frac{8}{9}\right)$. Then each of the remaining four intervals $\left[0, \frac{1}{9}\right], \left[\frac{2}{9}, \frac{1}{3}\right], \left[\frac{2}{3}, \frac{7}{9}\right]$ and $\left[\frac{8}{9}, 1\right]$ is divided into three and the central open intervals removed. The Cantor set is made of the points obtained after repeating this procedure an infinite number of times. By construction, this is the set of numbers made of 0 and 2 only, when expressed in the trinary form,

$$\{x = 0.x_1x_2 \ldots x_n \ldots | x_n = 0, 2; n \geq 1\}.$$

This set has a one-to-one correspondence with the set of points in [0, 1] expressed in binary form,

$$\{y = 0.y_1y_2 \ldots y_n \ldots | y_n = 0, 1; n \geq 1\}.$$

(This correspondence is realized if 2 in the trinary form is made correspondent with 1 in the binary form or *vice versa*, as $(0.220200\ldots)_3$ corresponds to

[2] It is a hypothesis (the continuum hypothesis) that there is no intermediate potency between those of a countable set and the continuum.

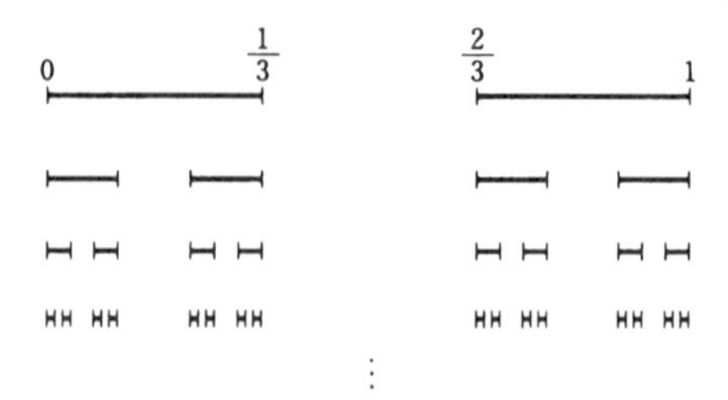

Figure 2B.3. The construction of the Cantor set.

(0.110100...) for example.) Thus the Cantor set is a continuum. Note however that this set has no length (i.e., measure zero, see appendix 2D). This is because the total length of the intervals removed is

$$\frac{1}{3}+2\left(\frac{1}{3}\right)^2+2^2\left(\frac{1}{3}\right)^3+\ldots=\frac{1}{3}\left\{1+\frac{2}{3}+\left(\frac{2}{3}\right)^2+\ldots\right\}=\frac{1}{3}\frac{1}{1-\frac{2}{3}}=1.$$

2C Upper limit and lower limit

The upper limit and the lower limit replace the rôle of the limit of the sequence when it does not exist. Let us consider an example first. The sequence $\left\{\sin^2\frac{n\pi}{6}+\frac{1}{n}\right\}$ takes values close to $0, \frac{1}{4}, \frac{3}{4}, 1$ in order when n is large and 'converges' to these four values as $n \to \infty$. Although this sequence has no limit, a subsequence may converge to one of four values. For example, the sequence $a_6, a_{12}, a_{18}, \ldots, a_{6n}, \ldots$ converges to 0, where $a_n = \sin^2\frac{n\pi}{6}+\frac{1}{n}$. Such values are called the *accumulation points* of the original sequence $\{a_n\}$. Moreover, the largest accumulation point is called the *upper limit* while the smallest is called the *lower limit* and is denoted by $\overline{\lim}_{n\to\infty} a_n$ and $\underline{\lim}_{n\to\infty} a_n$ or $\limsup_{n\to\infty} a_n$ and $\liminf_{n\to\infty} a_n$, respectively. In the example above, one has $\overline{\lim}_{n\to\infty} a_n = 1$ and $\underline{\lim}_{n\to\infty} a_n = 0$. These limits are identical if there is only one accumulation point, in which case the sequence has a limit. Examples of the upper limit and lower limit are given below.

(i) $a_n = (-1)^n + \dfrac{1}{2^n}$ $\quad \left(\overline{\lim}_{n\to\infty} a_n = 1, \quad \underline{\lim}_{n\to\infty} a_n = -1\right).$

(ii) $a_n = \sin^2\dfrac{n\pi}{4} + \dfrac{1}{n}$ $\quad \left(\overline{\lim}_{n\to\infty} a_n = 1, \quad \underline{\lim}_{n\to\infty} a_n = 0\right).$

(iii) An orbit $\{x_n\}$ of the tent map (1.3) or the binary transformation (1.4) with almost all irrational numbers as the initial value.

$\left(\overline{\lim}_{n\to\infty} x_n = 1, \quad \underline{\lim}_{n\to\infty} x_n = 0\right).$

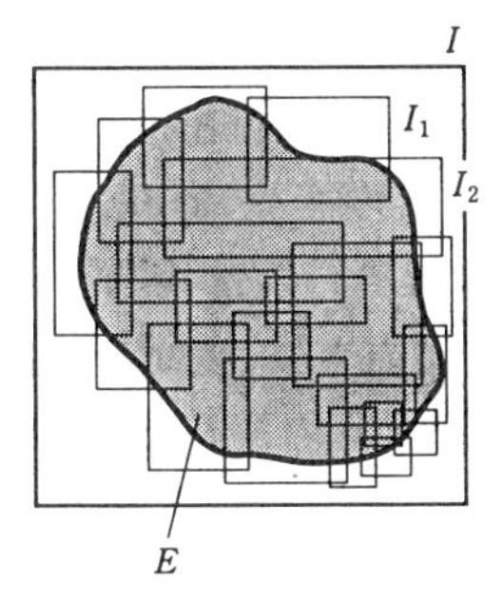

Figure 2D.1.

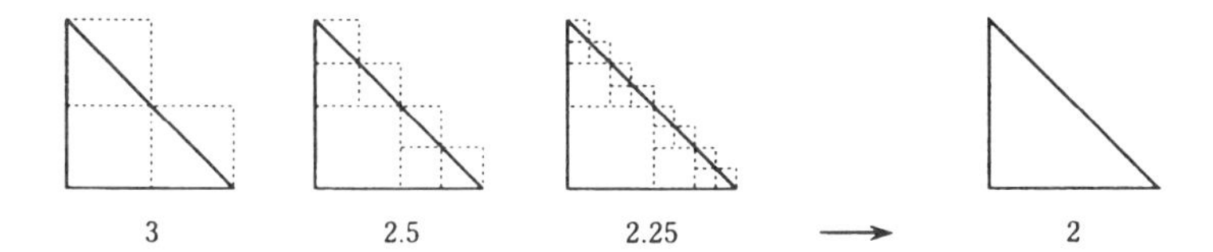

Figure 2D.2. An isosceles right-angled triangle whose shorter edges have length 2.

2D Lebesgue measure

Let us consider the Lebesgue measure of a set in a plane (planar figure) first to facilitate the geometrical image, since the Lebesgue measure generalizes the idea of length, area and volume.

Suppose a set E contained in a square I is covered with at most a countable (i.e., either finite or countably infinite) number of rectangles $I_1, I_2, \ldots, I_n, \ldots$ as shown in figure 2D.1. The total of the area $|I_n|$ of these rectangles, $\sum_{n\geq 1}|I_n|$, may depend on how E is covered. The infimum[3] of $\sum_{n\geq 1}|I_n|$ with respect to all possible coverings is called the Lebesgue *outer measure* and is denoted by $\mu^*(E)$:

$$\mu^*(E) = \inf \sum_{n\geq 1} |I_n|.$$

The area of an ordinary figure agrees with the outer measure (in fact, the outer measure in this case is just the measure) (see figure 2D.2).

The Lebesgue *inner measure* of a set E is introduced next. In contrast with

[3] *Infimum* is the generalization of the minimum value in a set of real numbers E. If a set has the minimum, the infimum of the set is the same as the minimum. Suppose a set E has no minimum. Then 0 in $\{1, \frac{1}{2}, \frac{1}{3}, \ldots, \frac{1}{n}, \ldots\}$ or -1 in the interval $(-1, 1)$, for example, is called the infimum of the respective set and is denoted by $\inf E$. Similarly the *supremum* is the generalization of the maximum and is denoted by $\sup E$. For example, $\sup(-1, 1) = 1$.

the outer measure, which measures the area from outside and hence is always greater than the area, the inner measure counts the area from inside and hence is smaller than the area. The outer measure of the complement $I - E$ of E with respect to I is $\mu^*(I - E)$. The Lebesgue inner measure is defined by subtracting this outer measure from the area $|I|$ of the square and is denoted by $\mu_*(E)$:

$$\mu_*(E) = |I| - \mu^*(I - E).$$

If the outer measure of a set E agrees with the inner measure, the set is called *Lebesgue measurable* and $\mu(E)$ $(= \mu^*(E) = \mu_*(E))$ is called the *Lebesgue measure*. The name Lebesgue (a French mathematician H Lebesgue (1875–1941)) is often omitted and the terms outer measure, inner measure, measurable and measure are used unless other measures are under consideration.

An ordinary figure is measurable and its measure is equal to its area. One has $\mu_*(E) < \mu^*(E)$ for an unmeasurable set E. To construct such an unmeasurable set, one employs the 'axiom of choice' of the set theory, the result of which is rather difficult to visualize.

A set with vanishing measure is called a *null set*. A null set may be said to have vanishing outer measure since

$$0 \leq \mu_*(E) \leq \mu^*(E)$$

in general. For example, a countable set is a null set. Let $E = \{x_1, x_2, \ldots, x_n, \ldots\}$ be a countable set and cover each point x_n with a square I_n with area $< \frac{\varepsilon}{2^n}$ with an arbitrary positive number ε $(n = 1, 2, \ldots)$. It is obvious that $E \subset \cup_{n=1}^{\infty} I_n$ and hence E is covered with these squares. Therefore it follows that

$$\sum_{n=1}^{\infty} |I_n| < \sum_{n=1}^{\infty} \frac{\varepsilon}{2^n} = \varepsilon \sum_{n=1}^{\infty} \frac{1}{2^n} = \varepsilon$$

namely $\mu^*(E) < \varepsilon$. Since ε may be arbitrarily small, $\mu^*(E)$ must vanish.

One may similarly define the measure of a subset of a real line (=the set of the real numbers) by considering intervals, instead of rectangles. For a solid object, i.e., a three-dimensional set, one employes rectangular parallelpipeds to define the measure.

The Cantor set is a typical example of a null set that is uncountable among subsets of the real line.

2E Normal numbers

Suppose a number $x \in [0, 1]$ is expressed as $x = (0.x_1x_2 \ldots x_n \ldots)_2$ as a binary number. Let $N(x, 0, n)$ and $N(x, 1, n)$ be the number of 0s and 1s, respectively, down to n decimal places in x. Similarly let $N(x, 00, n)$, $N(x, 01, n)$, $N(x, 10, n)$ and $N(x, 11, n)$ be the number of 00s, 01s, 10s and 11s in $0.x_1x_2 \ldots x_n$. (For example, given $x = (0.0100011\ldots)_2$, one finds $N(x, 00, 7) = 2$, $N(x, 01, 7) = 2$, $N(x, 10, 7) = 1$ and $N(x, 11, 7) = 1$.) In a similar way, let $N(x, a, n)$ be the number of patterns $a = a_1a_2 \ldots a_k$ of 0s and 1s in $x = 0.x_1x_2 \ldots x_n \ldots$ down to n decimal places. Then, if

$$\lim_{n\to\infty} \frac{1}{n} N(x, a, n) = \frac{1}{2^k}$$

is satisfied for any k (≥ 1) and for any pattern a of k 0s and 1s (there are 2^k such patterns), such a binary fraction x is called a binary *normal number*. Decimal normal numbers and r-nary normal numbers in general may be defined similarly. Clearly a rational number cannot be a normal number since it is a cyclic fraction. This concept was introduced by É Borel around 1900. He proved that almost all numbers in [0, 1] are r-nary normal numbers for an arbitrary r (≥ 2).

Although such a number as

$$(0.\,\underline{0}\;\underline{1}\;\underline{00}\;\underline{01}\;\underline{10}\;\underline{11}\;\underline{000}\;\underline{001}\;\underline{010}\;\underline{011}\;\underline{100}\ldots)_2$$

is a binary normal number, it is obviously not random. A random number[4] must be a normal number but the converse is not true. None of such 'natural' irrational numbers as $\sqrt{2}$, e and π has been proved to be a normal number. Needless to say, it is not clear whether they are random numbers or not. There are many examples of irrational numbers that are not normal.

Example 1. An example of a decimal fraction that is not a decimal normal number: a fraction in which a specific number does not appear.

Example 2. An example of a ternary fraction that is not a ternary normal number: an irrational number that is an element of the Cantor set in appendix 2B. The number is made of 0 and 2 only.

Example 3. An example of a binary fraction that is not a binary normal number: consider a number

$$(0.1\;\underbrace{0}_{1}\;1\;\underbrace{00}_{2}\;1\;\underbrace{000}_{3}\;1\;\underbrace{0000}_{4}\;10\ldots)_2$$

Since the number of 0s between two 1s gradually increases, it is not a cyclic fraction but an irrational number. Such a number is not a binary normal number since the distributions of 0s and 1s are different.

2F Periodic orbits with finite fraction initial value

Let us consider period 3 orbits of the tent map as an example. There are two of them: $\left(\frac{2}{7}, \frac{4}{7}, \frac{6}{7}\right)$ and $\left(\frac{2}{9}, \frac{4}{9}, \frac{8}{9}\right)$. Suppose one plots these periodic orbits with a computer. It is impossible to input the initial values $\frac{2}{7}$ and $\frac{2}{9}$ exactly.

[4] Although a random number has no rule in the arrangement of numbers, it may have statistical rules. Being a normal number is one such rule.

Instead, these numbers are rounded to finite fractions as $\frac{2}{7} = 0.285\,714$ and $\frac{2}{9} = 0.222\,222$. The input data must be finite fractions.

An odd-looking theorem, 'there is only one periodic orbit of $T(x)$, whose initial value is a finite fraction down to N decimal places for any N' is considered in the following [16].

Theorem. Let N be an arbitrary natural number. Then all the periodic orbits with initial values of the form $x_0 = 0.a_0a_1 \ldots a_{N-1}$ ($a_{N-1} \neq 0, 5$) are identical with the period $2 \times 5^{N-1}$.

Proof. Let us consider this problem for the binary transformation $B(x)$ first. Let $x_0 = 0.1 = \frac{1}{10}$ for example. The the orbit of B starting from x_0 is

$$\frac{1}{10},\ \underline{\frac{1}{5},\ \frac{2}{5},\ \frac{4}{5},\ \frac{3}{5}}\ ,\frac{1}{5},\ \frac{2}{5},\ \cdots$$

and hence it has a period 4. It is easily seen that any orbit of B with the initial condition given by a fraction down to one decimal place, except 0.5, is a period 4 orbit made up of the four points above. Next consider an orbit with the initial condition specified by a fraction down to two decimal places, $x_0 = 0.23$ for example. The orbit is

$$\frac{23}{100},\ \frac{23}{50},\ \underline{\frac{23}{25},\ \frac{21}{25},\ \frac{17}{25},\ \frac{9}{25},\ \frac{18}{25},\ \frac{11}{25},\ \frac{22}{25},\ \frac{19}{25},\ \frac{13}{25},\ \frac{1}{25},\ \frac{2}{25},}$$

$$\underline{\frac{4}{25},\ \frac{18}{25},\ \frac{16}{25},\ \frac{7}{25},\ \frac{14}{25},\ \frac{3}{25},\ \frac{6}{25},\ \frac{12}{25},\ \frac{24}{25}}\ ,\frac{23}{25},\ \cdots$$

which is periodic with period 20. These twenty period 20 points are of the form $\frac{k}{25}$ where k is not a multiple of 5. It can be shown that other orbits of $B(x)$ with the initial value of the form $x_0 = 0.a_0a_1$ ($a_1 \neq 0, 5$) converge to a period 20 orbit, whose 20 points are the same as the above orbit except that these points may be shifted in order (the reader should verify this statement by himself). This may be generalized as

'an orbit of $B(x)$ with the initial value $x_0 = 0.a_0a_1 \ldots a_{N-1}$ ($a_{N-1} \neq 0, 5$) converges to a period $4 \times 5^{N-1}$ orbit with $4 \times 5^{N-1}$ points of the form $k/5^N$ where k is not a multiple of 5.'

The point of the proof is that an orbit starting from

$$x_0 = 0.a_0a_1 \ldots a_{N-1} = \left(10^{N-1}a_0 + 10^{N-2}a_1 + \ldots + a_{N-1}\right)/10^N$$

eventually hits a number of the form $k/5^N$, after which the orbit takes values of this form only. If p is the period of such an orbit, p is the smallest number that satisfies $2^p k/5^N \equiv k/5^N \pmod 1$, namely $(2^p - 1)k/5^N \equiv 0 \pmod 1$. Since k is not a multiple of 5, one must have $2^p \equiv 1 \pmod{5^N}$. Our final observation is that the elementary number theory tells us that $p = \phi(5^N) = 4 \times 5^{N-1}$ ($\phi(n)$

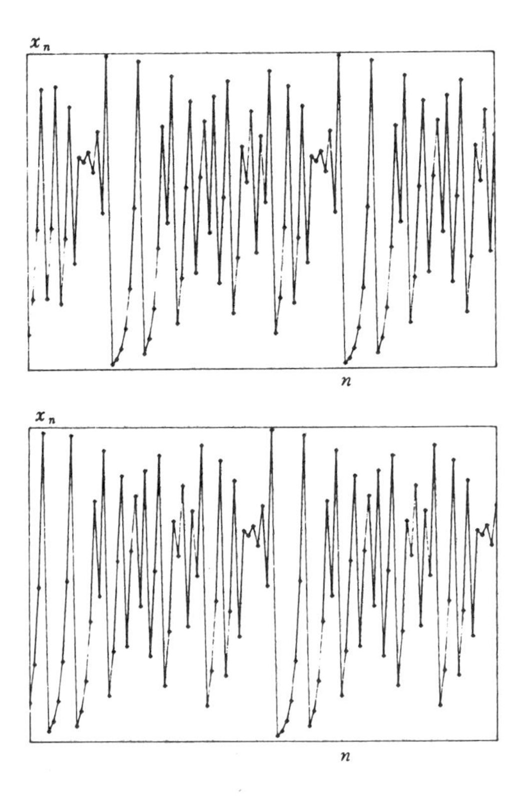

Figure 2F.1. Two orbits of the tent map with the initial conditions $x_0 = 0.111$ (above) and $x_0 = 0.123$ (below) up to x_{100}. The period is 50 and it is shown that they converge to the common periodic orbit.

is the Euler function, that is, the number of natural numbers less than n that are mutually prime to n) and that there are just $4 \times 5^{N-1}$ irreducible fractions of the form $k/5^N$ $(1 \le k < 5^N)$. Accordingly there is only one periodic orbit of B starting from a decimal fraction down to N decimal places.

Let us consider the tent map $T(x)$ next. An orbit of T with the initial value $x_0 = 0.1$ is

$$0.1 = \frac{1}{10},\ \frac{1}{5},\ \underline{\frac{2}{5},\ \frac{4}{5},}\ \frac{2}{5},\ \ldots$$

and hence it has period 2. When $x_0 = 0.23$, one finds

$$\frac{23}{100},\ \frac{23}{50},\ \frac{23}{25},\ \underline{\frac{4}{25},\ \frac{8}{25},\ \frac{16}{25},\ \frac{18}{25},\ \frac{14}{25},\ \frac{22}{25},\ \frac{6}{25},\ \frac{12}{25},\ \frac{24}{25},\ \frac{2}{25},}\ \frac{4}{25},\ \ldots$$

which shows that this orbit has period 10. These examples show that the periodic orbit of $T(x)$ with the initial condition $0.a_0a_1 \ldots a_{N-1}$ $(a_{N-1} \neq 0, 5)$ has a period

$2 \times 5^{N-1}$, which is just half of that for $B(x)$, and that these periodic points are of the form $2k/5^N$, where k is not a multiple of 5. The proof of these requires a detailed study of orbits of $B(x)$ and $T(x)$ with the common initial condition, which is omitted here. The uniqueness of the periodic orbit follows from the observations that an orbit of $T(x)$ with the initial value $x_0 = 0.a_0a_1 \ldots a_{N-1}$ eventually reaches a number of the form $2k/5^N$, k being not a multiple of 5, and that there are $2 \times 5^{N-1}$ of them, which agrees with the period.

Figure 2F.1 shows two orbits of $T(x)$ starting from two decimal fractions with three digits, the periodic parts of which match perfectly if one of them is shifted and superposed on the other. We note *en passant* that $2 \times 5^N = 50$ for $N = 3$, while there are about 1.1×10^{15} periodic orbits of $T(x)$ with period 50. Therefore only one of them has a finite decimal fraction as an initial condition.

2G The delta function

P A M Dirac considered the delta function $\delta(x)$, which satisfies

$$\int_a^b f(x)\delta(x - x_0)\,\mathrm{d}x = f(x_0) \qquad (x_0 \in [a, b]) \tag{2G.1}$$

for a continuous function $f(x)$. The 'function' $\delta(x - x_0)$ has a value only at $x = x_0$, as seen from the above definition, and there are an infinite number of them, so far as they satisfy equation (2G.1). Consider, for example, a function $\delta_\varepsilon(x - x_0)$ centred at x_0 with the width ε and the height $1/\varepsilon$ and take the limit $\varepsilon \to 0$. It follows from the intermediate-value theorem on integration that $\delta_\varepsilon(x)$ satisfies

$$\int_a^b f(x)\delta_\varepsilon(x - x_0)\,\mathrm{d}x = f\left(x_0 - \frac{\varepsilon}{2} + \varepsilon\theta(\varepsilon)\right). \tag{2G.2}$$

Here $0 \le \theta(\varepsilon) \le 1$ and $\left[x_0 - \frac{\varepsilon}{2}, x_0 + \frac{\varepsilon}{2}\right] \subset [a, b]$.

Since the r.h.s of equation (2G.2) is $f(x_0)$ in the limit of $\varepsilon \to 0$, we write

$$\lim_{\varepsilon \to 0} \delta_\varepsilon(x - x_0) = \delta(x - x_0).$$

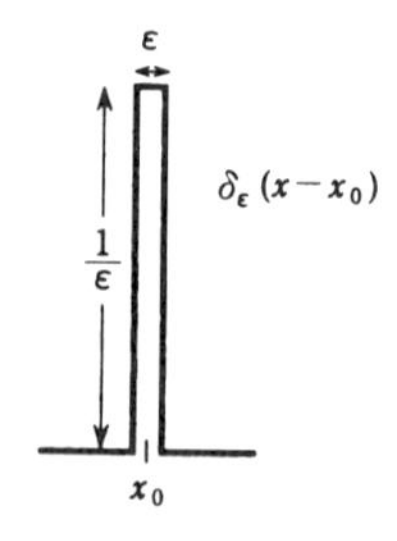

Figure 2G.1.

By taking $f(x) = 1$, one obtains

$$\int_a^b \delta(x - x_0)\, \mathrm{d}x = 1.$$

Frequently employed expressions of $\delta(x)$ are

$$\delta(x) = \lim_{n\to\infty} \sqrt{\frac{n}{\pi}}\, \mathrm{e}^{-nx^2} \qquad \delta(x) = \lim_{n\to\infty} \frac{\sin nx}{\pi x}.$$

Both of them take infinite value at $x = 0$ and they are considered to idealize various kinds of impulse or pulse appearing in engineering and physics. This viewpoint should be compared with the mathematical definition (2G.1).

A formal partial integration yields

$$\begin{aligned}\int_a^b f(x)\delta'(x - x_0)\, \mathrm{d}x &= \left[f(x)\delta(x - x_0)\right]_a^b - \int_a^b f'(x)\delta(x - x_0)\, \mathrm{d}x \\ &= -f'(x_0)\end{aligned}$$

which define the derivative of the δ-function.

The Heaviside function introduced in chapter 4,

$$H(x - x_0) = \begin{cases} 0 & (x < x_0) \\ 1 & (x \geq x_0) \end{cases}$$

can be expressed in terms of the δ-function as

$$H(x - x_0) = \int^x \delta(x' - x_0)\, \mathrm{d}x'.$$

3A Where does the period 3 window begin in the logistic map?

It will be shown that

$$R = 1 + \sqrt{8} = 3.828\,427\,125\ldots$$

is the beginning point of the window in the map $L_R(x) = Rx(1 - x)$. This is the case in which $L_R^3(x)$ is tangent with the line with unit gradient at three points, that is, the equation $L_R^3(x) - x = 0$ has three pairs of degenerate roots as shown in figure 3A.1. From the facts that there are other solutions $x = 0$ and that $L_R(x) - x$ has a solution $x = 1 - \frac{1}{R}$, one factorizes $L_R^3(x) - x$ as

$$\begin{aligned}L_R^3(x) - x = &-x(Rx + 1 - R)\{R^6x^6 - (3R^6 + R^5)x^5 \\ &+ (3R^6 + 4R^5 + R^4)x^4 - (R^6 + 5R^5 + 3R^4 + R^3)x^3 \\ &+ (2R^5 + 3R^4 + 3R^3 + R^2)x^2 - (R^4 + 2R^3 + 2R^2 + R)x \\ &+ (R^2 + R + 1)\}.\end{aligned} \tag{3A.1}$$

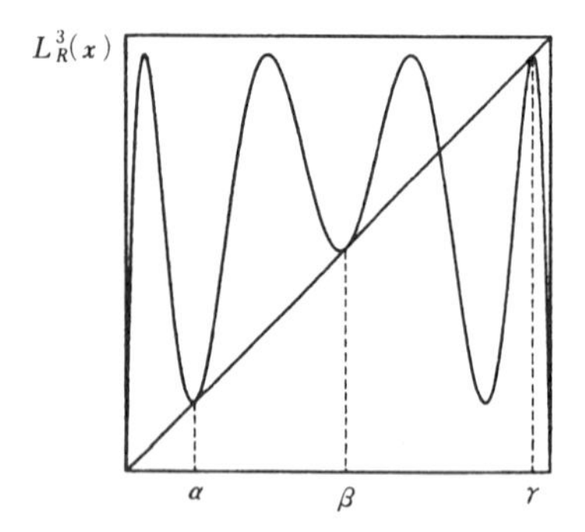

Figure 3A.1.

Here { } in the above equation must be written in terms of three double roots α, β and γ as

$$\{\ \} = R^6(x-\alpha)^2(x-\beta)^2(x-\gamma)^2. \tag{3A.2}$$

By putting

$$A = \alpha+\beta+\gamma \qquad B = \alpha\beta+\beta\gamma+\gamma\alpha \qquad C = \alpha\beta\gamma$$

one has

$$\begin{aligned}(x-\alpha)^2(x-\beta)^2(x-\gamma)^2 &= (x^3 - Ax^2 + Bx - C)^2 \\ &= x^6 - 2Ax^5 + (A^2+2B)x^4 - 2(AB+C)x^3 \\ &\quad + (B^2+2AC)x^2 - 2BCx + C^2. \end{aligned} \tag{3A.3}$$

Comparing this result with equation (3A.1), one obtains

$$\begin{aligned} A &= \frac{3R+1}{2R} \\ A^2+2B &= \frac{3R^2+4R+1}{R^2} \\ AB+C &= \frac{R^3+5R^2+3R+1}{2R^3} \\ B^2+2AC &= \frac{2R^3+3R^2+3R+1}{R^4} \\ BC &= \frac{R^3+2R^2+2R+1}{2R^5} \\ C^2 &= \frac{R^2+R+1}{R^6}. \end{aligned} \tag{3A.4}$$

The first three equations of (3A.4) determine A, B and C as

$$A = \frac{3R+1}{2R} \quad B = \frac{(3R+1)(R+3)}{8R^2} \quad C = \frac{-R^3+7R^2+5R+5}{16R^3}. \tag{3A.5}$$

By substituting these expressions into the rest of (3A.4), one obtains

$$\frac{(-3R^2+6R+5)(R^2-2R-7)}{64R^4} = 0$$

$$\frac{(R^3-5R^2-7R-7)(R^2-2R-7)}{64R^5} = 0$$

$$\frac{(R^4-12R^3+22R^2+20R+33)(R^2-2R-7)}{256R^6} = 0$$

from which one finds that the solutions of $R^2-2R-7=0$ are common solutions of the three equations above. By taking the inequality $0 < R \leq 4$ into account, one finally obtains $R = 1+\sqrt{8}$. The values of A, B and C are

$$A = \frac{20+\sqrt{8}}{14} = \frac{10+\sqrt{2}}{7} = 1.630\,601\,937\ldots$$

$$B = \frac{13+16\sqrt{2}}{49} = 0.727\,090\,1428\ldots$$

$$C = \frac{41\sqrt{2}-31}{343} = 0.078\,666\,93\ldots.$$

The cubic equation $x^3 - Ax^2 + Bx - C = 0$ is solved with the above values to yield

$$\alpha = 0.159\,928\,8183\ldots$$

$$\beta = 0.514\,355\,2768\ldots$$

$$\gamma = 0.956\,317\,819\ldots.$$

3B Newton method

Roots of an equation may be obtained iteratively by the Newton method. Let x_n be an approximate root in the nth step of iterations. The approximate root x_{n+1} in the next step is the intersection of the line

$$y - f(x_n) = f'(x_n)(x - x_n)$$

with the x-axis. Explicitly, x_{n+1} is given by

$$x_{n+1} - x_n = -\frac{f(x_n)}{f'(x_n)}. \tag{3B.1}$$

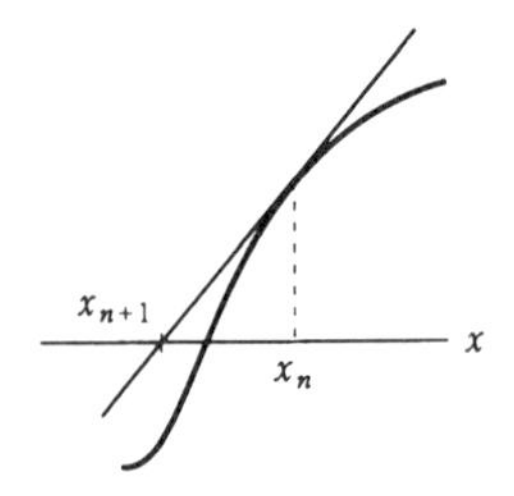

Figure 3B.1.

Since the Taylor expansion of $f(x)$ at $x = x_n$ is

$$f(x) = f(x_n) + (x - x_n)f'(x_n) + \dots$$

x_{n+1} thus obtained is the approximate solution $f(x_{n+1}) = 0$ of the linearized form above.

The Mathematica program to obtain the solution of $x^2 - 2 = 0$ by the Newton method is[5]

```
x=1.; dx=10^(-4); e=10;
While[Abs[e]>10^(-6), f=x^2-2.; g=(x-dx)^2-2.; e=-dx f/(f-g); x=x+e]
x
```

or simply

```
FindRoot[x^2-2==0, {x, 1}]
```

The program that gives the solution of $f(x) = 0$ starting from $x = x0$ is

```
FindRoot[f(x)==0, {x, x0}]
```

where one gives $f(x)$ and $x0$ explicitly.

As an application, let us obtain the parameter R of the superstable periodic orbit of the logistic map with period N. The Mathematica program is

```
fn[R_, n_]:=Module[{x, nx}, x[0]=1/2;
x[nx_]:=R x[nx-1] (1-x[nx-1]); x[n]-1/2]
FindRoot[fn[R, 5]==0, {R, 3.99}]
```

for example. A proper choice of R0 (=3.99 in the above example) and n=5 yield

$$3.738\,92\dots \qquad 3.905\,71\dots \qquad 3.990\,27\dots$$

and so on.

In the old days when no electronic calculators with functions were available (younger readers might not imagine this!), the mth root of A was obtained by hand using the Newton method as follows.

[5] The original text gives (the Japanese version of) the BASIC program. This and the following Mathematica programs are provided by the translator.

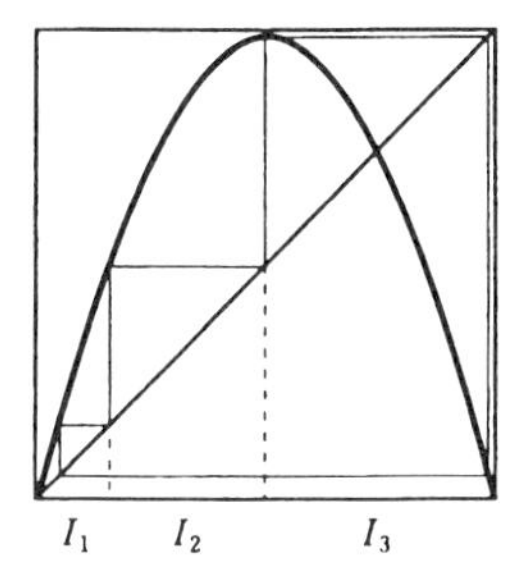

Figure 3C.1.

Choose $f(x) = x^m - A$, which yields $f'(x) = mx^{m-1}$. One obtains, from equation (3B.1), that

$$x_{n+1} = x_n - \frac{x_n^m - A}{mx_n^{m-1}} = \frac{1}{m}\left\{(m-1)x_n + \frac{A}{x_n^{m-1}}\right\}.$$

For example, take $m = 3$ and $A = 2$ to find

$$x_{n+1} = \frac{1}{3}\left(2x_n + \frac{2}{x_n{}^2}\right).$$

The choice $x_1 = 1$ produces

$$x_2 = 1.3333\ldots \qquad x_3 = 1.2643\ldots \qquad x_4 = 1.2599\ldots.$$

The exact value is $\sqrt[3]{2} = 1.259\,921\,05\ldots$.

3C How to evaluate topological entropy

The method of the structure matrix[6] is employed for superstable periodic orbits. Let us consider, as an example, the superstable period 4 point $R = 3.960\,270\ldots$ of the logistic map $L_R(x)$ shown in figure 3C.1. Take intervals I_1, I_2 and I_3 as in figure 3C.1. Since

$$L_R(I_1) = I_2 \qquad L_R(I_2) = I_3 \qquad L_R(I_3) = I_1 + I_2 + I_3$$

it is convenient to define the following matrix M, called the *structure matrix*:

$$M = \begin{pmatrix} L_R(I_1) \\ L_R(I_2) \\ L_R(I_3) \end{pmatrix} \begin{pmatrix} 0 & 1 & 0 \\ 0 & 0 & 1 \\ 1 & 1 & 1 \end{pmatrix} \begin{pmatrix} I_1 \\ I_2 \\ I_3 \end{pmatrix}.$$

[6] See [17] for details.

The largest eigenvalue of M, denoted by λ_{max}, is

$$\lambda_{max} = 1.839\,28\ldots \qquad 1.722\,08\ldots.$$

Then the topological entropy h is given by

$$h = \log \lambda_{max}.$$

The same method yields λ_{max} of the superstable orbit with period 5 shown in figure 3.21 as $1.512\,87\ldots$, $1.722\,08\ldots$ and $1.927\,56\ldots$.

The evaluation of the topological entropy from the superstable periodic orbit is applicable only for limited values of R in practice, even when these orbits are dense in R.

In contrast, the *kneading sequence* method, explained below, may be applicable to cases with an arbitrary R.

Let us take $x_0 = 1/2$ as the initial value and consider the ith iteration. Let

$$K_i = \begin{cases} R & (f^i(1/2) > 1/2) \\ C & (f^i(1/2) = 1/2) \\ L & (f^i(1/2) < 1/2) \end{cases}$$

and construct the kneading sequence

$$\boldsymbol{K}_N(f) = K_1 K_2 K_3 \ldots K_N.$$

Define ε_i, using these K_i, as

$$\varepsilon_i = \begin{cases} +1 & (K_i = L) \\ -1 & (K_i = R). \end{cases}$$

By repeating this, one also defines

$$\varepsilon_k = \prod_{i=1}^{k-1} \varepsilon_i \quad (K_k = C).$$

Then a polynomial $\boldsymbol{P}_{K.N}(\tau)$ of order N in τ is defined as

$$\boldsymbol{P}_{K.N}(\tau) = 1 + \sum_{n=0}^{N} \left(\prod_{j=1}^{n} \varepsilon_j \right) \tau^n. \qquad (3C.1)$$

The smallest solution to the equation

$$\lim_{N\to\infty} \boldsymbol{P}_{K.N}(\tau) = 0 \qquad (3C.2)$$

yields the topological entropy as[7]

$$h(f) = -\log\tau.$$

The limit $N \to \infty$ cannot be taken in equation (3C.2), in practice, and one is obliged to use equation (3C.1) with finite N. There are a small number of terms to be kept in the case where τ is small. If, in contrast, τ is close to unity, tens of terms are not sufficient. The 'regula falsi' is more appropriate than the Newton method to find roots of a polynomial with large order.

3D Examples of invariant measure

Examples of invariant measures, defined in chapter 2, are given for the logistic map

$$x_{n+1} = Rx_n(1 - x_n)$$

for several choices of R.

The invariant measure $\rho(x)$ is given by equation (2.12). There are infinite kinds of $\rho(x)$ depending upon the choice of the initial value. What is observed in a computer simulation is the 'natural' invariant measure covering the largest number of points. This is the measure corresponding to the stationary orbits of a map introduced in chapters 2 and 3. The orbits of the logistic map are shown in figure 3D.1 for comparison.

4A Generalized dimension D_q is monotonically decreasing in q

Consider the generalized dimension

$$D_q = \frac{1}{q-1}\lim_{l\to 0}\frac{\log\sum_{i=1}^{n(l)} p_i^q}{\log l} = \lim_{l\to 0}\frac{\log\left(\sum_{i=0}^{n(l)} p_i^q\right)^{\frac{1}{q-1}}}{\log l}.$$

Since $\log l < 0$, one is required to show that

$$f(x) = \left(\sum p_i{}^{x+1}\right)^{\frac{1}{x}}$$

is monotonically increasing in $x \equiv q - 1$.

The following inequality is employed to show this. Let $p > q > 0$ and $a_i, b_i > 0$ $(1 \le i \le n)$. Then Hölder's inequality

$$\left(\frac{a_1 b_1^q + a_2 b_2^q + \ldots + a_n b_n^q}{a_1 + a_2 + \ldots + a_n}\right)^{\frac{1}{q}} \le \left(\frac{a_1 b_1^p + a_2 b_2^p + \ldots + a_n b_n^p}{a_1 + a_2 + \ldots + a_n}\right)^{\frac{1}{p}} \qquad (4A.1)$$

[7] See Collet P, Crutchfield J P and Eckmann J P 1983 *Commun. Math. Phys.* **88** 257 and Block L, Keesling J, Li S and Peterson K 1989 *J. Stat. Phys.* **55** 929.

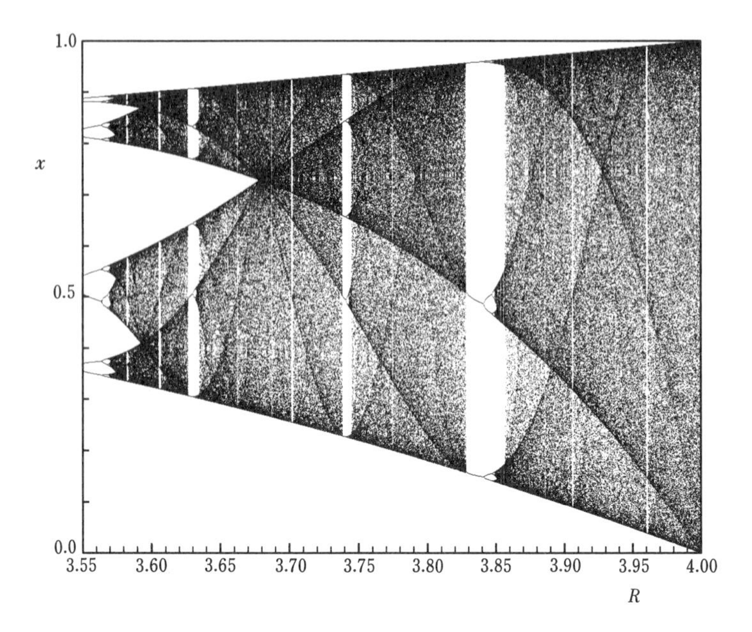

Figure 3D.1. Stationary orbits of the logistic map.

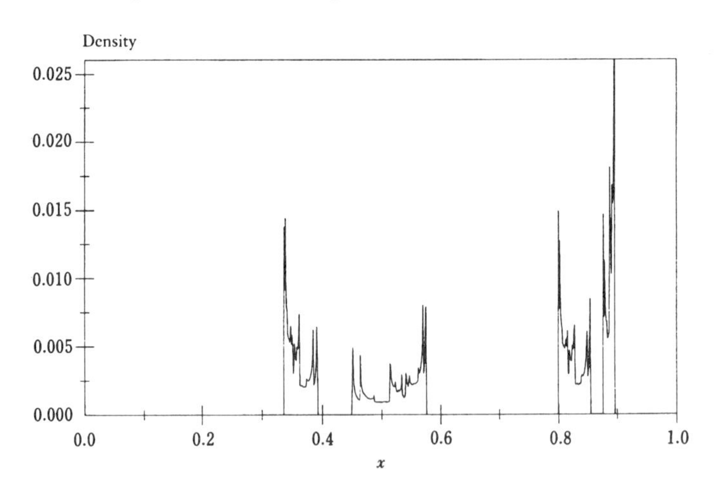

Figure 3D.2. $R = 3.58$. Right above the beginning of chaos, where $R_\infty = 3.569\,9456\ldots$. Chaos here has four bands. The invariant measure of this and the following figures are obtained after 5×10^6 iterations. The ordinate shows the density.

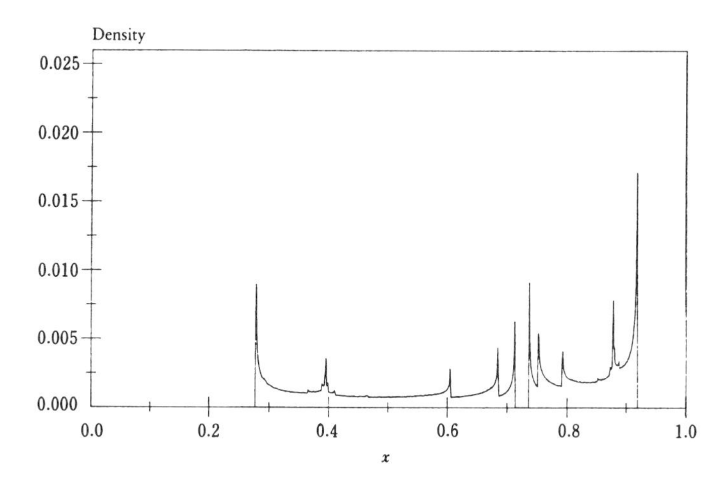

Figure 3D.3. $R = 3.67$. Chaos with two bands. This is the invariant measure obtained slightly below the window with period 10. The superstable periodic orbit in the window with period 10 exists at $R = 3.673\,0082\ldots$.

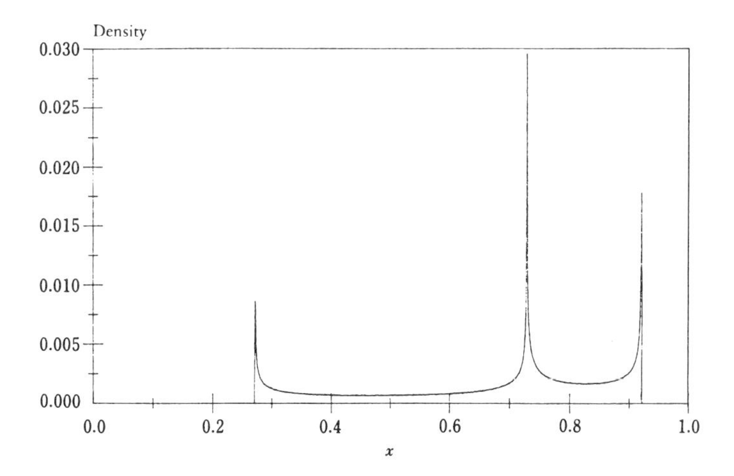

Figure 3D.4. $R = 3.6786$. This is an invariant measure when two bands coalesce. The precise value of R is $R = \frac{2}{3} + \frac{16}{3\sqrt[3]{152+24\sqrt{33}}} + \frac{\sqrt[3]{152+24\sqrt{33}}}{3} = 3.678\,5735\ldots$.

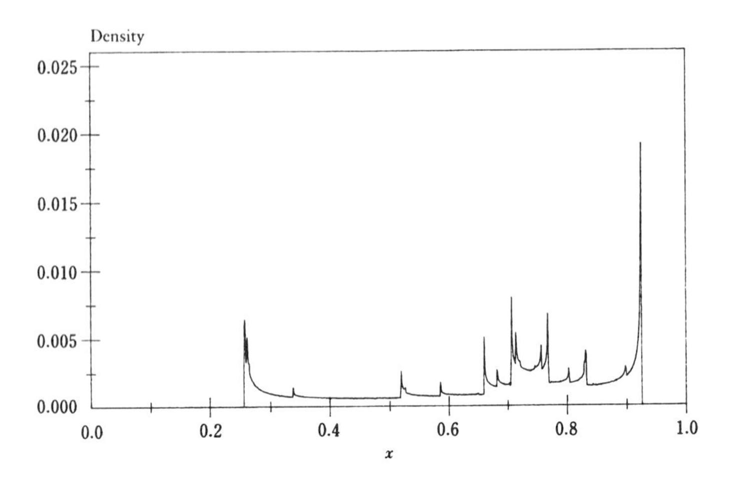

Figure 3D.5. $R = 3.70$. Chaos is made up of a single band.

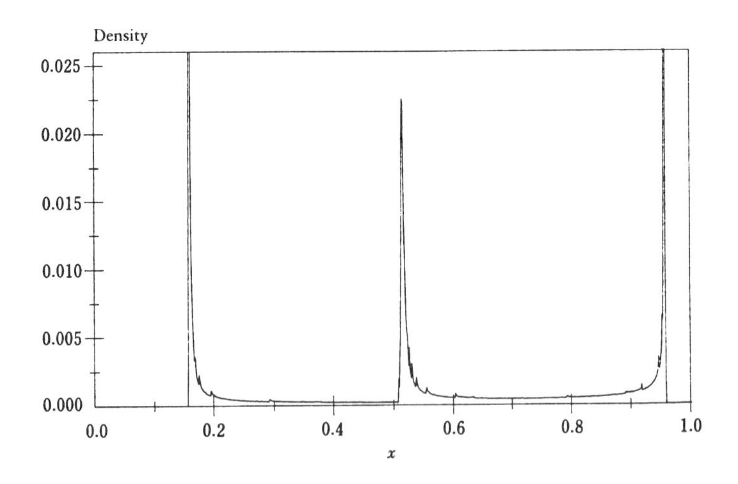

Figure 3D.6. $R = 3.8283$. The invariant measure which shows intermittent chaos at slightly below $R = 1 + \sqrt{8} = 3.828\,4271\ldots$, where the window with period 3 starts.

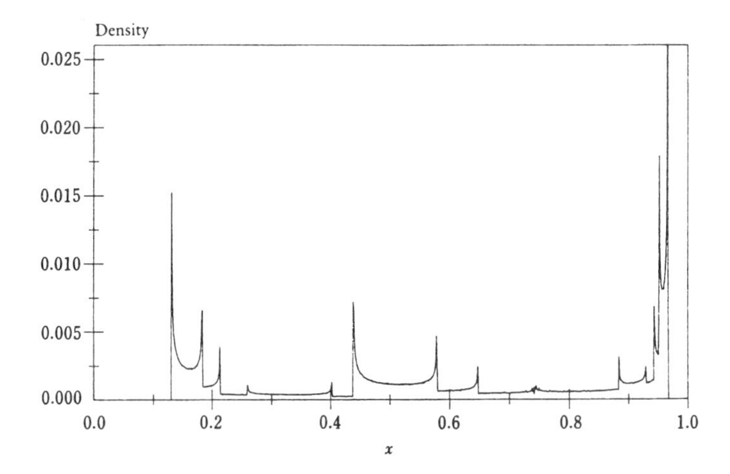

Figure 3D.7. $R = 3.86$. The invariant measure above the period 3 window. The influence of the growth of tiny period 3 chaos within the period 3 window is observed.

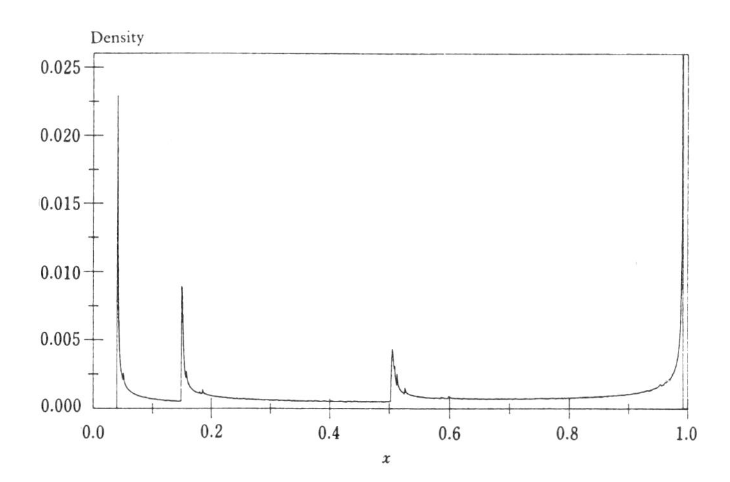

Figure 3D.8. $R = 3.96$. The invariant measure slightly below the period 4 window. The superstable period 4 orbit within the period 4 window exists at $R = 3.960\,270\,127\ldots$.

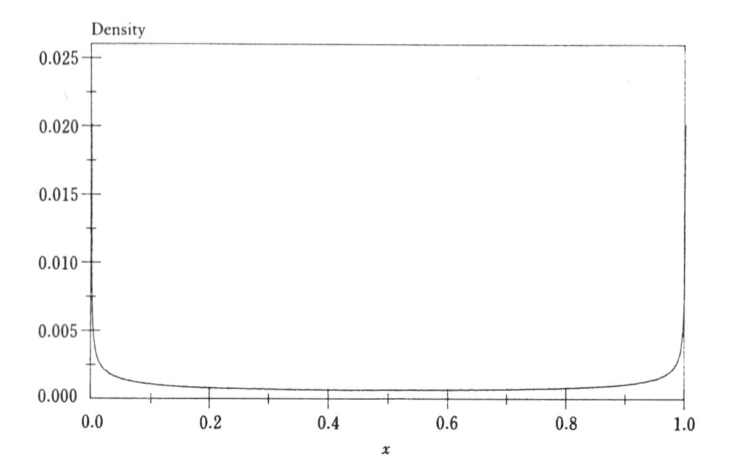

Figure 3D.9. The invariant measure at $R = 4$. The density is given by $\rho(x) = \frac{1}{\pi\sqrt{x(1-x)}}$ as mentioned in chapter 2.

is satisfied.

If one puts $a_i = b_i = p_i$, it follows from $\sum_{i=1}^{n} p_i = 1$ that

$$\left(\sum_{i=1}^{n} p_i{}^{q+1}\right)^{\frac{1}{q}} \leq \left(\sum_{i=1}^{n} p_i{}^{p+1}\right)^{\frac{1}{p}} \qquad (p > q > 0). \tag{4A.2}$$

This shows that $f(x)$ is monotonically increasing when $x > 0$.

Consider the case $x < 0$ next. If one puts $p^* = -p$ and $q^* = -q$ in equation (4A.1), one obtains

$$\left(\frac{\sum_{i=1}^{n} a_i b_i^{-q^*}}{\sum_{i=1}^{n} a_i}\right)^{-\frac{1}{q^*}} \leq \left(\frac{\sum_{i=1}^{n} a_i b_i^{-p^*}}{\sum_{i=1}^{n} a_i}\right)^{-\frac{1}{p^*}}$$

since $p^* < q^* < 0$. It follows from the above equation that

$$\left(\frac{\sum_{i=1}^{n} a_i b_i^{-p^*}}{\sum_{i=1}^{n} a_i}\right)^{\frac{1}{p^*}} \leq \left(\frac{\sum_{i=1}^{n} a_i b_i^{-q^*}}{\sum_{i=1}^{n} a_i}\right)^{\frac{1}{q^*}}. \tag{4A.3}$$

If p^*, q^*, b_i^{-1} are written as p, q, b_i, respectively, it follows from $p < q < 0$

and $a_i, b_i > 0$ $(1 \le i \le n)$ that

$$\left(\frac{\sum_{i=1}^{n} a_i b_i^p}{\sum_{i=1}^{n} a_i}\right)^{\frac{1}{p}} \le \left(\frac{\sum_{i=1}^{n} a_i b_i^q}{\sum_{i=1}^{n} a_i}\right)^{\frac{1}{q}}$$

which shows that an inequality of the form (4A.1) is satisfied in the case $p < q < 0$ as well. Accordingly, $f(x)$ is monotonically increasing when $x < 0$. Note also that

$$\lim_{x\to 0} \log f(x) = \sum p_i \log p_i$$

and hence

$$\lim_{x\to 0+} f(x) = \lim_{x\to 0-} f(x).$$

Therefore, $f(x)$ is monotonically increasing for $-\infty < x < \infty$.

4B Saddle point method

Let $f(x)$ and $g(x)$ be real-valued functions and consider the integral of the form

$$I = \int_{-\infty}^{+\infty} f(x) \mathrm{e}^{-g(x)}\, \mathrm{d}x.$$

The function $g(x)$ is assumed to take a minimum value at $x = x_0$, namely

$$g'(x_0) = 0 \qquad g''(x_0) > 0.$$

The Taylor expansion of $g(x)$ around $x = x_0$ is

$$g(x) = g(x_0) + \tfrac{1}{2}(x - x_0)^2 g''(x_0) + \dots.$$

Keeping the second order term in $(x - x_0)$ in the above expansion, the above integral is approximated as

$$\begin{aligned} I &\simeq \int_{-\infty}^{+\infty} f(x)\, \mathrm{e}^{-\{g(x_0)+\frac{1}{2}(x-x_0)^2 g''(x_0)\}}\, \mathrm{d}x \\ &= \mathrm{e}^{-g(x_0)} \sqrt{\frac{2}{g''(x_0)}} \int_{-\infty}^{+\infty} f\left(x_0 + \sqrt{\frac{2}{g''(x_0)}}\xi\right) \mathrm{e}^{-\xi^2}\, \mathrm{d}\xi \end{aligned}$$

where $\xi = \sqrt{g''(x_0)/2}(x - x_0)$.

The term $e^{-\xi^2}$ rapidly approaches zero as $|\xi|$ increases. If the variation of $f(x_0 + \sqrt{2}\xi)$ is sufficiently slow compared to this, it follows from

$$\int_{-\infty}^{+\infty} e^{-\xi^2} d\xi = \sqrt{\pi}$$

that the approximate value of the integral is

$$\begin{aligned} I &\simeq \sqrt{\frac{2}{g''(x_0)}} f(x_0) e^{-g(x_0)} \int_{-\infty}^{+\infty} e^{-\xi^2} d\xi \\ &= \sqrt{\frac{2\pi}{g''(x_0)}} f(x_0) e^{-g(x_0)}. \end{aligned}$$

4C Chaos in double pendulum

A pendulum with its string made of a spring, defined by equation (4.23), is a system showing chaos. A double pendulum is superior to a pendulum with a spring, however, since the former is easier to construct and chaos is more easily observable in the former. Figure 1.16 shows an actual double pendulum we have constructed. An irregular motion may be observed before the oscillation dies out if ball bearings are used in its axes and if the support is concrete enough that the propagation of the oscillation of the system to the table is negligible.

The equations of motion of this system are more complicated than those of a pendulum with a spring (see equation (4.23)). The analysis of this system is easily done by employing the Lagrangian. If the reader is unfamiliar with this formalism, he should be referred to a textbook on mechanics.

The potential energy P of the system shown in figure 4C.1 is

$$\begin{aligned} P &= m_1 l_1 g(1 - \cos\theta_1) + m_2\{l_1(1 - \cos\theta_1) + l_2(1 - \cos\theta_2)\}g \\ &= (m_1 + m_2)l_1 g(1 - \cos\theta_1) + m_2 l_2 g(1 - \cos\theta_2). \end{aligned} \quad (4C.1)$$

Let us introduce the variables

$$x_1 = l_1 \sin\theta_1 \qquad y_1 = l_1(1 - \cos\theta_1)$$

$$x_2 = l_1 \sin\theta_1 + l_2 \sin\theta_2 \qquad y_2 = l_1(1 - \cos\theta_1) + l_2(1 - \cos\theta_2)$$

to write down the kinetic energy K as

$$\begin{aligned} K &= \tfrac{1}{2}m_1(\dot{x}_1^2 + \dot{y}_1^2) + \tfrac{1}{2}m_2(\dot{x}_2^2 + \dot{y}_2^2) \\ &= \tfrac{1}{2}m_1 l_1^2\dot{\theta}_1^2 + \tfrac{1}{2}m_2\{l_1^2\dot{\theta}_1^2 + l_2^2\dot{\theta}_2^2 + 2l_1 l_2\dot{\theta}_1\dot{\theta}_2\cos(\theta_1 - \theta_2)\}. \end{aligned} \quad (4C.2)$$

Then construct the Lagrangian $L = K - P$ to obtain the Euler–Lagrange

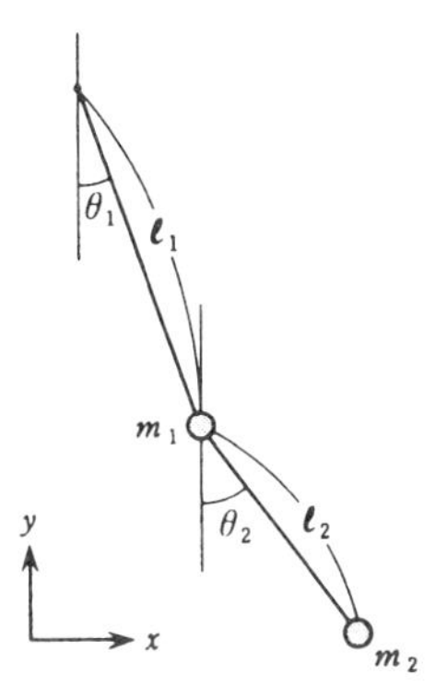

Figure 4C.1. The parameters and the variables of a double pendulum.

equations

$$\begin{aligned}\frac{\mathrm{d}}{\mathrm{d}t}\left(\frac{\partial L}{\partial\dot{\theta}_1}\right)-\frac{\partial L}{\partial\theta_1}&=0\\ \frac{\mathrm{d}}{\mathrm{d}t}\left(\frac{\partial L}{\partial\dot{\theta}_2}\right)-\frac{\partial L}{\partial\theta_2}&=0\end{aligned}\tag{4C.3}$$

which take the following explicit forms

$$\begin{aligned}\dot{\alpha}_1+\mu_2 l_{21}\dot{\alpha}_2\cos(\theta_1-\theta_2)+\mu_2 l_{21}\alpha_2^2\sin(\theta_1-\theta_2)+\omega_1^2\sin\theta_1&=0\\ l_{21}\dot{\alpha}_2+\dot{\alpha}_1\cos(\theta_1-\theta_2)-\alpha_1^2\sin(\theta_1-\theta_2)+\omega_2^2\sin\theta_2&=0\end{aligned}\tag{4C.4}$$

where $\alpha_1=\dot{\theta}_1,\ \alpha_2=\dot{\theta}_2, \mu_2=m_2/(m_1+m_2), l_{21}=l_2/l_1$ and $\omega_1^2=g/l_1$.

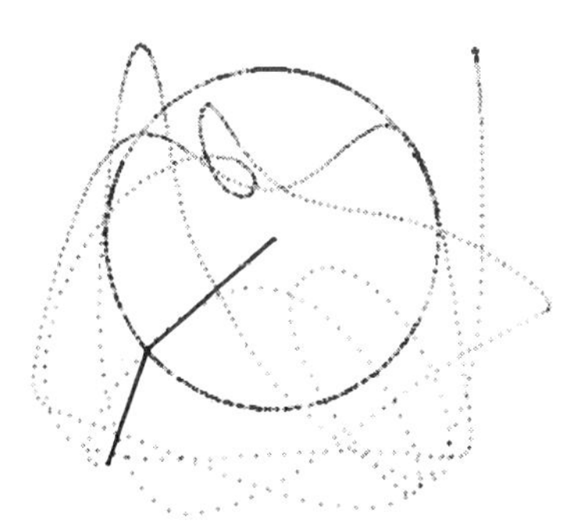

Figure 4C.2. The motion of a double pendulum obtained using the given program. The mass m_1 is shown in white circles while m_2 in black circles.

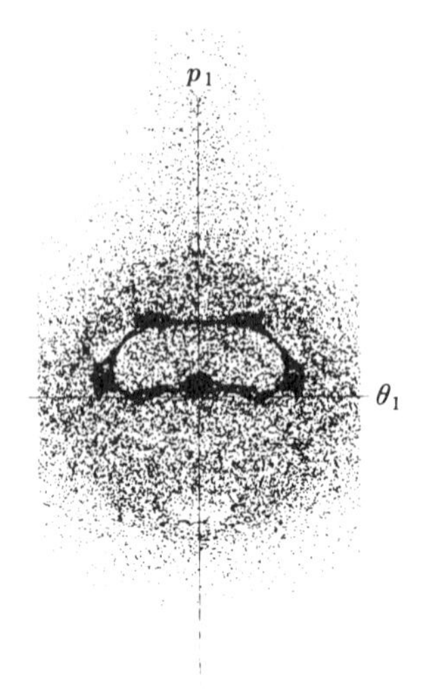

Figure 4C.3. The Poincaré section of a double pendulum. The points lie in the plane $\theta_2 = 0$ in the space $(p_1, \theta_1, \theta_2)$. The distribution of the points shows that the motion is chaotic. The initial conditions are: $\theta_1 = 2\pi/3, \theta_2 = 5\pi/6$ and $\dot{\theta}_1 = \dot{\theta}_2 = 0$. The variable p_1 is the conjugate momentum of θ_1.

A Mathematica program which solves these equations is shown below with the result.

```
l21=1/Sqrt[2.]; mu2=.6; om1=1.; om2=Sqrt[1/l21] om1;
s2=Sqrt[2.]; pi=N[Pi, 10];

sol=NDSolve[{th1''[t]+mu2 l21 th2''[t] Cos[th1[t]-th2[t]]+
    mu2 l21 (th2'[t])^2 Sin[th1[t]-th2[t]]+om1^2 Sin[th1[t]]==0,
    l21 th2''[t]+th1''[t] Cos[th1[t]-th2[t]]-
    (th1'[t])^2 Sin[th1[t]-th2[t]]+om2^2 Sin[th2[t]]==0,
    th1[0]==2 pi/3, th2[0]==2 pi/3, th1'[0]==1, th2'[0]==0},
    {th1, th2}, {t, 0, 20.1}, MaxSteps->2000];

p1[t_]:={s2 Cos[th1[t]], s2 Sin[th1[t]]}
p2[t_]:={s2 Cos[th1[t]]+Cos[th2[t]], s2 Sin[th1[t]]+Sin[th2[t]]}

pt1=Table[Graphics[{Thickness[0.0001], Circle[p1[t], 0.04]},
    PlotRange->{{-2.5, 2.5}, {-2.5, 2.5}},
    AspectRatio->1]/. sol, {t, 0, 20, 0.05}];
pt2=Table[Graphics[Disk[p2[t], 0.04],
    PlotRange->{{-2.5, 2.5}, {-2.5, 2.5}},
    AspectRatio->1]/. sol, {t, 0, 20, 0.05}];
```

```
Show[pt1, pt2]
```

Large values for `th1[0]` and `th2[0]` ($>2\pi/3$) generate sufficiently irregular motion. The initial velocities are specified by `th1'[0]` and `th2'[0]`. It is found from the Poincaré section in figure 4C.3 that this motion is chaotic.

4D Singular points and limit cycle of van der Pol equation

The van der Pol equation (4.19) is transformed into simultaneous differential equations

$$\begin{cases} \dot{x} = y \\ \dot{y} = -x + \varepsilon(1 - x^2)y \end{cases} \tag{4D.1}$$

by introducing $y = \dot{x}$. The singular point, $\dot{x} = \dot{y} = 0$, is the origin $(0, 0)$ of the xy-plane. We will be concerned with the behaviour of the solution near this singular point and the existence of the limit cycle.

Let $P(x, y)$ and $Q(x, y)$ be smooth functions which satisfy $P(0, 0) = Q(0, 0) = 0$. Consider the behaviour of the solution of the autonomous system

$$\begin{cases} \dot{x} = P(x, y) \\ \dot{y} = Q(x, y) \end{cases} \quad \text{or} \quad \frac{\mathrm{d}y}{\mathrm{d}x} = \frac{Q(x, y)}{P(x, y)} \tag{4D.2}$$

near the origin. By writing the partial derivatives of P and Q at the origin as $A = P_x(0, 0)$, $B = P_y(0, 0)$, $C = Q_x(0, 0)$ and $D = Q_y(0, 0)$, the behaviour of equation (4D.2) near the origin is approximately written as

$$\begin{cases} \dot{x} = Ax + By \\ \dot{y} = Cx + Dy \end{cases} \quad \text{or} \quad \frac{\mathrm{d}y}{\mathrm{d}x} = \frac{Cx + Dy}{Ax + By}. \tag{4D.3}$$

Then the behaviour of the solution of equation (4D.3) in the neighbourhood of the origin is classified by the matrix

$$\Delta = \begin{pmatrix} A & B \\ C & D \end{pmatrix}$$

assuming $|\Delta| = AD - BC \neq 0$. Let λ_1 and λ_2 be the solutions of the eigenvalue equation

$$|\Delta - \lambda I| = \begin{vmatrix} A - \lambda & B \\ C & D - \lambda \end{vmatrix} = \lambda^2 - (A + D)\lambda + |\Delta| = 0.$$

Then the solutions are classified as follows:

(1) If λ_1 and λ_2 are different real solutions and $|\Delta| > 0$, the origin is a nodal point.

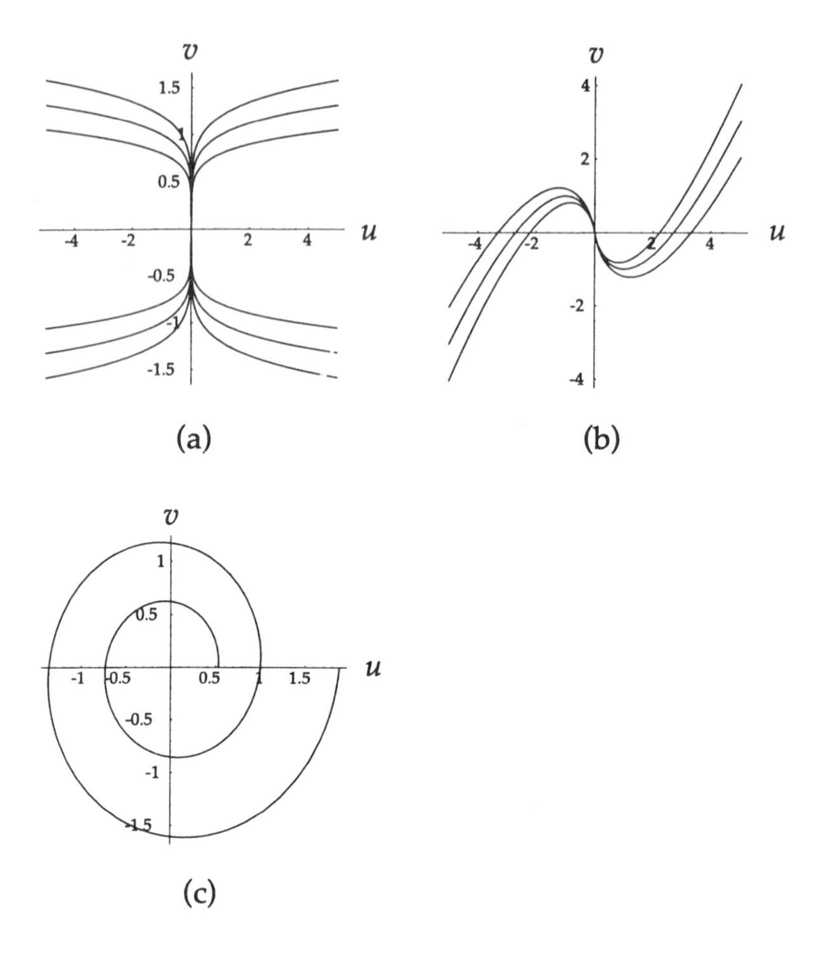

Figure 4D.1. Solutions of the van der Pol equation with (a) $\varepsilon = 2\sqrt{2}$ and $\lambda_2/\lambda_1 = (\sqrt{2}-1)/(\sqrt{2}+1) = 0.1715\ldots$. (b) $\varepsilon = 2$ and (c) $\varepsilon = 0.2$, $\alpha = 0.1$, $\beta = \sqrt{1-0.01} \simeq 1$ and hence $\alpha/\beta \simeq 0.1$.

(2) If λ_1 and λ_2 are different real solutions and $|\Delta| < 0$, the origin is a saddle point.
(3) If $\lambda_1 = \lambda_2$ are degenerate solutions, the origin is a nodal point.
(4) If λ_1 and λ_2 are complex solutions and $A + D \neq 0$, the origin is a spiral point.
(5) If λ_1 and λ_2 are complex solutions and $A + D = 0$, the origin is a focal point.

We only consider the cases (1), (3) and (4) since only these cases are relevant for the van der Pol equation.

Since

$$\Delta = \begin{pmatrix} 0 & 1 \\ -1 & \varepsilon \end{pmatrix} \qquad |\Delta| = 1 \qquad |\Delta - \lambda I| = \lambda^2 - \varepsilon\lambda + 1 = 0$$

for the van der Pol equation, there appear the following three cases according to the magnitude of the positive constant ε:

(1) $\varepsilon > 2$. The eigenvalues

$$\lambda_1 = \frac{\varepsilon}{2} + \sqrt{\left(\frac{\varepsilon}{2}\right)^2 - 1} \qquad \lambda_2 = \frac{\varepsilon}{2} - \sqrt{\left(\frac{\varepsilon}{2}\right)^2 - 1}$$

are both positive. Introduce a new set of variables $\begin{pmatrix} u \\ v \end{pmatrix}$ by applying the linear transformation

$$\begin{pmatrix} C & \lambda_1 - A \\ C & \lambda_2 - A \end{pmatrix} = \begin{pmatrix} -1 & \lambda_1 \\ -1 & \lambda_2 \end{pmatrix}$$

obtained from the corresponding eigenvectors on the old variables $\begin{pmatrix} x \\ y \end{pmatrix}$. Then equation (4D.3) is transformed into

$$\begin{cases} \dot{u} = \lambda_1 u \\ \dot{v} = \lambda_2 v \end{cases} \quad \text{or} \quad \frac{\mathrm{d}v}{\mathrm{d}u} = \frac{\lambda_2}{\lambda_1}\frac{v}{u}.$$

The solution of the above equation is $v = \text{constant}|u|^{\lambda_2/\lambda_1}$, whose graph is shown in figure 4D.1(a). Although the graph is deformed in the xy-plane, its qualitative behaviour remains unchanged.

(2) $\varepsilon = 2$. The eivengalues are degenerate, $\lambda_1 = \lambda_2 = 1$. Equation (4D.3) becomes

$$\begin{cases} \dot{u} = u \\ \dot{v} = u + v \end{cases} \quad \text{or} \quad \frac{\mathrm{d}v}{\mathrm{d}u} = \frac{u+v}{u}$$

under the linear transformation

$$\begin{pmatrix} C & -\frac{1}{2}(A - D) \\ 0 & \frac{1}{2}(A + D) \end{pmatrix} = \begin{pmatrix} -1 & 1 \\ 0 & 1 \end{pmatrix}.$$

The solution of this equation is $v = u(\log|u| + \text{constant})$, whose graph is shown in figure 4D.1(b).

(3) $0 < \varepsilon < 2$. The eigenvalues are

$$\lambda_1 = \alpha + \mathrm{i}\beta \qquad \lambda_2 = \alpha - \mathrm{i}\beta \qquad \left(\alpha = \frac{\varepsilon}{2} \quad \beta = \sqrt{1 - \left(\frac{\varepsilon}{2}\right)^2}\right)$$

and equation (4D.3) becomes

$$\begin{cases} \dot{u} = au - v \\ \dot{v} = u + av \end{cases} \quad \text{or} \quad \frac{dv}{du} = \frac{u + av}{au - v}$$

under the linear transformation

$$\begin{pmatrix} C & \alpha - A \\ 0 & \beta \end{pmatrix} = \begin{pmatrix} -1 & \alpha \\ 0 & \beta \end{pmatrix}.$$

The solution of this equation is $r = \text{constant}\, e^{\alpha\theta/\beta}$ in the polar coordinates in the uv-plane. The solution spirals around the origin away to infinity as shown in figure 4D.1(c).

It is shown next that the solution of equation (4D.1) has a limit cycle when ε is small. Let $r^2 = x^2 + y^2$. Substituting $\dot{y} = -x + \varepsilon(1 - x^2)y$ into $r\dot{r} = x\dot{x} + y\dot{y}$, one obtains

$$r\dot{r} = \varepsilon(1 - x^2)y^2. \tag{4D.4}$$

Note that when $\varepsilon = 0$, (4D.1) becomes

$$\begin{cases} \dot{x} = y \\ \dot{y} = -x \end{cases} \tag{4D.5}$$

from which one obtains $\frac{dy}{dx} = -\frac{x}{y}$, that is, $x\,dx + y\,dy = 0$. The solution of this equation is $x^2 + y^2 = \text{constant}$. Thus the solution of equation (4D.5) is given by $x = r\sin t$, $y = r\cos t$, where r is a constant in the present case. In the presence of small $\varepsilon > 0$, x and y are perturbed slightly and become

$$x = r\sin t + \delta_1 \qquad y = r\cos t + \delta_2.$$

Substituting these into equation (4D.4) and neglecting infinitesimal quantities one obtains

$$r\dot{r} = \varepsilon(1 - r^2\sin^2 t)r^2\cos^2 t \qquad \dot{r} = \varepsilon(1 - r^2\sin^2 t)r\cos^2 t.$$

The variation Δr of r while t increases from 0 to 2π is

$$\begin{aligned} \Delta r &= \int dr = \int_0^{2\pi} \frac{dr}{dt}\,dt \\ &= \varepsilon r \int_0^{2\pi} (1 - r^2\sin^2 t)\cos^2 t\,dt \\ &= \varepsilon\pi r\left(1 - \frac{r^2}{4}\right) \end{aligned}$$

where r is assumed to be almost constant. Therefore it turns out that the circle $r = 2$ is a limit cycle since $r > 2$ leads to $\Delta r < 0$ while $r < 2$ to $\Delta r > 0$: a solution starting from a point inside the circle winds itself round the circle from inside while that starting from outside winds itself from outside. Note however that the circle is deformed as shown in figure 4.4 when ε is not infinitesimal. The origin remains a spiral point (figure 4D.1(c)), even in this case, and a solution which starts off at a point near the origin spirals away from the origin and winds round the limit cycle from inside.

4E Singular points of the Rössler model

Let us consider the singular points of the Rössler model

$$\begin{cases} \dot{x} = -y - z \\ \dot{y} = x + ay \\ \dot{z} = b + z(x - \mu) \end{cases} \tag{4E.1}$$

where a, b and μ are positive constants. The conditions

$$y + z = 0 \qquad x + ay = 0 \qquad b + z(x - \mu) = 0$$

are obtained from $\dot{x} = \dot{y} = \dot{z} = 0$. The above simultaneous quadratic equations may be solved and two singular points are obtained. When $a = b = \frac{1}{5}$, in particular, the singular points are $(\alpha, -5\alpha, 5\alpha)$ and $(\beta, -5\beta, 5\beta)$, where

$$\alpha = \frac{5\mu - \sqrt{25\mu^2 - 4}}{10} \qquad \beta = \frac{5\mu + \sqrt{25\mu^2 - 4}}{10}.$$

Note that α and β are solutions of a quadratic equation $5x^2 - 5\mu x + \frac{1}{5} = 0$.

Let us consider the behaviour of a solution in the neighbourhood of a singular point $(\alpha, -5\alpha, \alpha)$. Let us put

$$x = \alpha + p \qquad y = -5\alpha + q \qquad z = 5\alpha + r$$

to this end and substitute them into equation (4E.1) to obtain

$$\begin{cases} \dot{p} = -q - r \\ \dot{q} = p + \dfrac{1}{5} q \\ \dot{r} = 5\alpha p + (\alpha - \mu) r \end{cases} \tag{4E.2}$$

where the second order infinitesimal quantity pr has been ignored. If one puts $\mu = 5.7$ here, $\alpha = 0.007\,02\ldots$ is obtained. Then the term with α may be dropped from equation (4E.2) and the last equation simplifies as $\dot{r} = -\mu r$, from which one obtains $r = A\,\mathrm{e}^{-\mu t}$. If follows from this solution that $r \to 0$ as

$t \to \infty$, namely $z \to 5\alpha = 0.035\,13\ldots$. As for p and q, one assumes the forms $p = B\,\mathrm{e}^{\lambda t}$ and $q = C\,\mathrm{e}^{\lambda t}$ and substitutes them into the first two equations in equation (4E.2) to obtain

$$\begin{cases} \dot{p} = \lambda p = -q - r \\ \dot{q} = \lambda q = p + \frac{1}{5}q. \end{cases}$$

Since the variable r is negligibly small when t is large, the above equations reduce to

$$\begin{cases} -\lambda p - q = 0 \\ p + \left(\frac{1}{5} - \lambda\right) q = 0 \end{cases}$$

Then one must have

$$\begin{vmatrix} -\lambda & -1 \\ 1 & \frac{1}{5} - \lambda \end{vmatrix} = \lambda^2 - \frac{1}{5}\lambda + 1 = 0$$

from which one obtains the solutions $\lambda = 0.1 \pm \sqrt{0.99}\mathrm{i} \simeq 0.1 \pm 0.995\mathrm{i}$. Therefore p and q are written as

$$p = C_1\,\mathrm{e}^{0.1t}\sin(0.995t + \theta_1) \qquad q = C_2\,\mathrm{e}^{0.1t}\sin(0.995t + \theta_2)$$

where C_1, C_2, θ_1 and θ_2 are constants. As $t \to \infty$, the solution spirals away from the centre $(\alpha, -5\alpha, 5\alpha)$ with period $\sim 2\pi$, practically staying on the plane $z = 5\alpha$.

References

[1] Li T Y and Yorke J A 1975 *Am. Math. Mon.* **82** 985

[2] Sharkovskii A N 1964 *Ukr. Mat. J.* **16** 61

[3] Devaney R L 1986 *An Introduction to Chaotic Dynamical Systems* (Reading, MA: Benjamin–Cummings)

[4] Smítal J 1986 *Trans. AMS* **297** 269

[5] Nathanson M V 1976 *Proc. AMS* **60** 75

[6] Kubo I 1989 *Tech. Rep. IEICE, NLP* **88** 58

[7] Baba Y, Kubo I and Takahashi Y 1996 *Nonlinear Anal. Theor. Methods Appl.* **26** 1611

[8] Smítal J 1983 *Proc. AMS* **87** 54

[9] Milnor J and Thurston W 1988 *On Iterated Maps of the Interval (Lecture Notes in Mathematics 1342)* (New York: Springer) p 465

[10] Ito S, Tanaka S and Nakada H 1979 *Tokyo J. Math.* **2** 221

[11] Feigenbaum M J 1979 *J. Stat. Phys.* **21** 669

[12] Takens F 1981 *Dynamical Systems and Turbulence (Warwick, 1980)* ed D Rand and L S Young (Berlin: Springer) p 366

[13] Mori H 1980 *Prog. Theor. Phys.* **63** 1044

[14] Sano M and Sawada Y 1985 *Phys. Rev. Lett.* **55** 1082

[15] Hardy G H and Wright E M 1959 *An Introduction to the Theory of Numbers* 4th edn (Oxford: Oxford University Press)

[16] Baba Y and Nagashima H 1989 *Prog. Theor. Phys.* **81** 541

[17] Takahashi Y 1980 Chaos and periodic points of interval dynamics *Seminar Rep. Math. Dept Tokyo Metropolitan University* (in Japanese)

Solutions

Chapter 1

Problem 1. Figure 1.5; the tent map. Figure 1.6; the logistic map. Figure 1.7; the Bernoulli shift.

Problem 2.

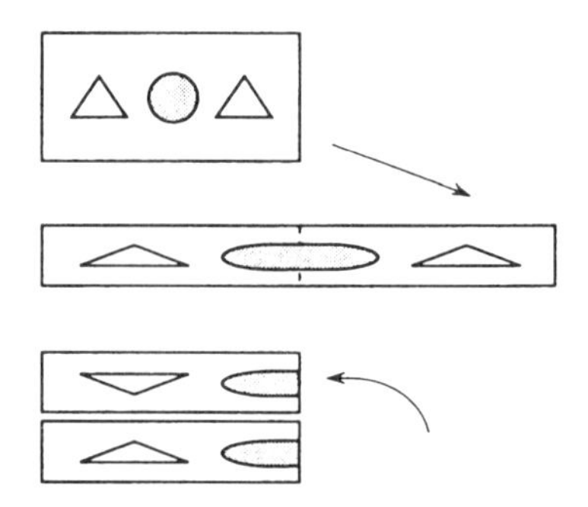

Problem 3. It is found from $10^{-10} \times 2^n = 1$ that $n = 33.2\ldots$, that is, 33 or 34 times.

Chapter 2

Problem 1. The number of period n orbits is $A(n)/n$ since each of the period n orbits are made of n period n points and these points do not belong to other periodic orbits (why?).

Problem 2. Binary expressions of 13 and 15 are $13 = (1101)_2$ and $15 = (1111)_2$. The fraction $13/15$ is obtained in the binary form as

```
          0.1101
     ___________
1111)1101.0000
      111 1
      ______
      101 10
       11 11
      ______
        1 1100
          1111
          ____
          1101
```

The remainder 1101 is equal to the number to be divided and hence the rest of the division is a repetition of the step taken so far. Therefore one finds $(0.\dot{1}10\dot{1})_2 = 13/15$. Conversely this can be checked as

$$(0.\dot{1}10\dot{1})_2 = 0.\underbrace{1101}\,\underbrace{1101}\ldots = \left(\frac{1}{2}+\frac{1}{4}+\frac{1}{16}\right)+\frac{1}{16}\left(\frac{1}{2}+\frac{1}{4}+\frac{1}{16}\right)+\ldots$$
$$= \frac{13}{16}\left(1+\frac{1}{16}+\frac{1}{16^2}+\ldots\right) = \frac{13}{16}\frac{1}{1-\dfrac{1}{16}} = \frac{13}{15}.$$

Now, a period 4 orbit of the binary transformation B starting from $13/15 = (0.\dot{1}10\dot{1})_2$ is $\{0.\dot{1}10\dot{1},\ 0.\dot{1}01\dot{1},\ 0.\dot{0}11\dot{1},\ 0.\dot{1}11\dot{0},\ 0.\dot{1}10\dot{1},\ldots\}$, since B is a shift transformation of binary numbers. This orbit is also expressed as $\left\{\frac{13}{15},\ \frac{11}{15},\ \frac{7}{15},\ \frac{14}{15},\ \frac{13}{15},\ldots\right\}$ in a fractional form.

Problem 3. If $x = x_0$ is substituted into

$$g'(x) = f'(f^{n-1}(x))f'(f^{n-2}(x))\ldots f'(f(x))f'(x)$$

one obtains

$$g'(x_0) = f'(x_{n-1})f'(x_{n-2})\ldots f'(x_1)f'(x_0).$$

Next, substitute $x = x_k$ $(1 \le k \le n-1)$ into $g'(x)$ above. Take $k = 1$ for example to find

$$g'(x_1) = f'(f^{n-1}(x_1))f'(f^{n-2}(x_1))\ldots f'(f(x_1))f'(x_1).$$

Since

$$f^{n-1}(x_1) = f^{n-1}(f(x_0)) = f^n(x_0) = x_0$$

(x_0 is a period n point of f!) and

$$f^{n-2}(x_1) = f^{n-1}(x_0) = x_{n-1}, \quad f^{n-3}(x_1) = f^{n-2}(x_0) = x_{n-2}, \ldots$$

one finds

$$g'(x_1) = f'(x_0)f'(x_{n-1})f'(x_{n-2})\ldots f'(x_2)f'(x_1).$$

Similarly, one obtains

$$g'(x_0) = g'(x_1) = \ldots = g'(x_{n-1}) = f'(x_0)f'(x_1)\ldots f'(x_{n-1}).$$

Thus the derivatives of $g(x) = f^n(x)$ at points $x_0, x_1, \ldots, x_{n-1}$ on a period n orbit are identical.

Problem 4. Put $x = x_0$ in $g'(x) = f'(f(x))f'(x)$, $g''(x) = f''(f(x))f'(x)^2 + f'(f(x))f''(x)$ to find

$$g'(x_0) = f'(f(x_0))f'(x_0) = f'(x_0)^2 = (-1)^2 = 1$$
$$g''(x_0) = f''(f(x_0))f'(x_0)^2 + f'(f(x_0))f''(x_0)$$
$$= f''(x_0)(f'(x_0)^2 + f'(x_0)) = f''(x_0)((-1)^2 + (-1)) = 0.$$

Problem 5. Let $\alpha = \{I_n\}_n$ be an open covering of I so that $\cup_n I_n \supset I$. It can be shown, then, that

$$f^{-1}\left(\bigcup_n I_n\right) = \bigcup_n f^{-1}(I_n)$$

in general. (This is true for any pair of a map f and a set with an arbitrary potency. See books on set theory.) It follows from the above identity and $f^{-1}(I) = I$ that

$$f^{-1}\left(\bigcup_n I_n\right) = \bigcup_n f^{-1}(I_n) \supset f^{-1}(I) = I.$$

Moreover, since f is a continuous map and $\{I_n\}$ is an open covering, it is found that $f^{-1}(I_n)$ is an open set. Accordingly, $f^{-1}\alpha$ is an open covering.

Next, let $\alpha = \{I_n\}_n$ and $\beta = \{J_m\}_m$ be open coverings of I. Then $I_n \cap J_m$ is an open set as an intersection of two open sets and clearly $\cup_{n,m} I_n \cap J_m \supset I$. Thus $\alpha \vee \beta$ is an open covering.

Problem 6.

$$f^{-3}\alpha = \left\{\left[0, \frac{1}{8}\right], \left[\frac{1}{8}, \frac{1}{4}\right], \left[\frac{1}{4}, \frac{1}{2}\right], \left[\frac{1}{2}, \frac{5}{8}\right], \left[\frac{5}{8}, \frac{11}{16}\right], \left[\frac{11}{16}, \frac{3}{4}\right], \left[\frac{3}{4}, \frac{7}{8}\right], \left[\frac{7}{8}, 1\right]\right\}.$$

Problem 7. If $\text{lap}(f^n) = A\alpha^n + B\beta^n + C$ $(A > 0,\ \alpha > 1,\ \alpha > |\beta| > 0)$, one finds

$$\begin{aligned} \log \text{lap}(f^n) &= \log\left[A\alpha^n \left(1 + \frac{B}{A}\left(\frac{\beta}{\alpha}\right)^n + \frac{C}{A}\frac{1}{\alpha^n}\right)\right] \\ &= n \log \alpha + \log A + \log\left(1 + \frac{B}{A}\left(\frac{\beta}{\alpha}\right)^n + \frac{C}{A}\frac{1}{\alpha^n}\right). \end{aligned}$$

Since $\log\left(1 + \frac{B}{A}\left(\frac{\beta}{\alpha}\right)^n + \frac{C}{A}\frac{1}{\alpha^n}\right) \to 0$ as $n \to \infty$, one finds

$$h(f) = \log \alpha.$$

Problem 8. Omitted.

Problem 9. Let $x = 0.x_1x_2\ldots$ be a binary normal number and let (a, b) be an arbitrary open interval contained in $[0, 1]$. Moreover, let I_k^i be a binary interval

$$\left(\frac{i}{2^k}, \frac{i+1}{2^k}\right) \quad (0 \le i \le 2^k - 1, k \ge 1).$$

Since x is a binary normal number, one obtains, for $0 \le i \le 2^k - 1$,

$$\lim_{n\to\infty} \frac{1}{n} N(x, I_k^i, n) = \frac{1}{2^k}.$$

Note here that $1/2^k = |I_k^i|$. If a and b are (binary) rational numbers, then they can be reduced to a common denominator;

$$a = \frac{i}{2^k} \qquad b = \frac{j}{2^k} \qquad (0 \le i < j \le 2^k).$$

Since $N(x, J, n) = \sum_{l=i}^{j-1} N(x, I_k^l, n)$ for a binary interval $J = \left(\frac{i}{2^k}, \frac{j}{2^k}\right)$, one obtains

$$\lim_{n\to\infty} \frac{1}{n} N(x, J, n) = \lim_{n\to\infty} \frac{1}{n} \sum_{l=i}^{j-1} N(x, I_k^l, n) = \sum_{l=i}^{j-1} \frac{1}{2^k} = \frac{j-i}{2^k} = |J|.$$

Any a and b may be approximated by binary rational numbers so that

$$\lim_{n\to\infty} \frac{1}{n} N(x, (a, b), n) = b - a.$$

Problem 10.

(i) *For* $0 \le y_n = \frac{2}{\pi}\theta_n \le \frac{1}{2}$. One has $y_{n+1} = 2y_n = \frac{4}{\pi}\theta_n$ in this case. One has $\theta_{n+1} = 2\theta_n$ since $y_{n+1} = \frac{2}{\pi}\theta_{n+1}$. It also follows from $\theta_n = \sin^{-1}\sqrt{x_n}$ that $x_n = \sin^2\theta_n$. Accordingly

$$\begin{aligned} x_{n+1} &= \sin^2\theta_{n+1} = \sin^2 2\theta_n = 4\sin^2\theta_n \cos^2\theta_n = 4\sin^2\theta_n\,(1 - \sin^2\theta_n) \\ &= 4x_n(1 - x_n). \end{aligned}$$

(ii) *For* $\frac{1}{2} \le y_n \le 1$. One has $y_{n+1} = 2 - 2y_n = 2 - \frac{4}{\pi}\theta_n = \frac{2}{\pi}\theta_{n+1}$, from which one finds $\theta_{n+1} = \frac{\pi}{2}\left(2 - \frac{4}{\pi}\theta_n\right) = \pi - 2\theta_n$. Accordingly

$$x_{n+1} = \sin^2\theta_{n+1} = \sin^2(\pi - 2\theta_n) = \sin^2 2\theta_n = 4x_n(1 - x_n).$$

Thus $x_{n+1} = 4x_n(1 - x_n) = L(x_n)$ in both cases.

Problem 11. Two sequences x_n and y_n are related by $x = \sin^2 \frac{\pi}{2} y$. It follows from $\frac{dx}{dy} = 2 \sin \frac{\pi}{2} y \cos \frac{\pi}{2} y \frac{\pi}{2} = \frac{\pi}{2} \sin \pi y$ that $\left| \frac{dx}{dy} \right| \leq \frac{\pi}{2}$. Thus

$$|x' - x| = \left| \sin^2 \frac{\pi}{2} y' - \sin^2 \frac{\pi}{2} y \right| \leq \frac{\pi}{2} |y' - y|.$$

Therefore, if the sequence $\{y_n\}$ is dense in $[0, 1]$, the corresponding sequence $\{x_n\}$ is also dense.

Problem 12. Suppose

$$f(x) = T(x) = 1 - |2x - 1| \qquad \rho(x) = 1 \qquad (0 \leq x \leq 1).$$

Then one finds

$$I = \int \delta(y - f(x))\rho(x)\mathrm{d}x = \int_0^{1/2} \delta(y - 2x)\mathrm{d}x + \int_{1/2}^1 \delta(y - 2 + 2x)\mathrm{d}x$$

for $0 \leq y \leq 1$. Since

$$\int_0^{1/2} \delta(y - 2x)\mathrm{d}x = \frac{1}{2} \int_0^1 \delta(y - u)\mathrm{d}u = \frac{1}{2} \qquad (u = 2x)$$

$$\int_{1/2}^1 \delta(y - 2 + 2x)\mathrm{d}x = \frac{1}{2} \int_0^1 \delta(y - u)\mathrm{d}u = \frac{1}{2} \qquad (u = -2x + 2)$$

which follow from

$$\int \delta(y - u)\mathrm{d}u = \int \delta(u - y)\mathrm{d}u = 1$$

one obtains

$$I = 1 = \rho(y).$$

Next, suppose

$$f(x) = 4x(1 - x) \qquad \rho(x) = \frac{1}{\pi\sqrt{x(1 - x)}} \qquad (0 < x < 1).$$

Then

$$I = \int \delta(y - f(x))\rho(x)\mathrm{d}x = \int_0^1 \delta(y - 4x(1 - x))\frac{\mathrm{d}x}{\pi\sqrt{x(1 - x)}}$$

for $0 < y < 1$. By putting $4x(1 - x) = u$, one finds

$$\begin{aligned} I &= \int_0^{1/2} \delta(y - 4x(1 - x))\frac{\mathrm{d}x}{\pi\sqrt{x(1 - x)}} + \int_{1/2}^1 \delta(y - 4x(1 - x))\frac{\mathrm{d}x}{\pi\sqrt{x(1 - x)}} \\ &= \int_0^1 \delta(y - u)\frac{\mathrm{d}u}{2\pi\sqrt{u(1 - u)}} + \int_0^1 \delta(y - u)\frac{\mathrm{d}u}{2\pi\sqrt{u(1 - u)}} \\ &= \frac{1}{\pi}\int_0^1 \delta(y - u)\frac{1}{\sqrt{u(1 - u)}}\mathrm{d}u = \frac{1}{\pi}\frac{1}{\sqrt{y(1 - y)}} = \rho(y). \end{aligned}$$

Problem 13. Put $\theta = \frac{\pi}{2} - \alpha$ in the integral

$$I = \frac{2}{\pi}\int_0^{\pi/2} \log(\cos\theta)\mathrm{d}\theta$$

to obtain

$$I = \frac{2}{\pi}\int_{\pi/2}^{0} \log(\sin\alpha)(-\mathrm{d}\alpha) = \frac{2}{\pi}\int_0^{\pi/2} \log(\sin\alpha)\mathrm{d}\alpha.$$

Then it follows that

$$\begin{aligned}
2I &= \frac{2}{\pi}\int_0^{\pi/2} \log(\cos\theta)\mathrm{d}\theta + \frac{2}{\pi}\int_0^{\pi/2} \log(\sin\theta)\mathrm{d}\theta \\
&= \frac{2}{\pi}\int_0^{\pi/2} \log\left(\frac{\sin 2\theta}{2}\right)\mathrm{d}\theta \\
&= \frac{2}{\pi}\int_0^{\pi/2} \log(\sin 2\theta)\mathrm{d}\theta - \frac{2}{\pi}\int_0^{\pi/2} (\log 2)\mathrm{d}\theta \\
&= \frac{2}{\pi}\int_0^{\pi} \{\log(\sin\theta')\}\frac{1}{2}\mathrm{d}\theta' - \log 2 = I - \log 2
\end{aligned}$$

where $\theta' = 2\theta$. Thus one proves

$$I = -\log 2.$$

Chapter 3

Problem 1. Any point starting from (0, 1) in the figure below must fall into the interval $[2A(1-A), A]$ and cannot escape from there.

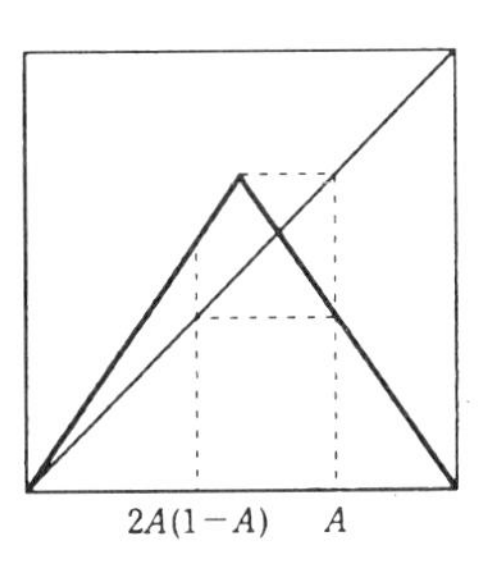

Problem 2. Since it follows from $|-R^2+2R+4| < 1$ that $(R^2-2R-5)(R^2-2R-3) < 0$, R satisfies $1-\sqrt{6} < R < -1$ and $3 < R < 1+\sqrt{6}$. Since R is

defined in $0 < R \leq 4$, one finds $3 < R < 1 + \sqrt{6}$.

Problem 3. The functions $L_R^4(x) - x$, $L_R^2(x) - x$ and $L_R(x) - x$ are polynomials of order 16, 4 and 2 in x, respectively. Since

$$L_R^4(x) - x = L_R^2(L_R^2(x)) - x$$

vanishes if $L_R^2(x) = x$ is satisfied, $L_R^4(x) - x$ has a factor $L_R^2(x) - x$ by the factor theorem. It is similarly shown that it has a factor $L_R(x) - x$.

Problem 4. Put $x = 0$ in equation (3.11):

$$g(0) = -\alpha g[g(0)].$$

This is to be confirmed for $g(x)$ of equation (3.9). First one finds $g(0) = 1$. On the other hand,

$$g(1) = 1 - 1.527\,63 + 0.104\,815 - 0.026\,7057 = -0.4495\ldots$$

and

$$-\alpha g(1) \simeq -2.50 \times (-0.45) \simeq 1.125$$

from which one finds a rough agreement within the expansion of $g(x)$.

Problem 5. It follows from $g\left(-\frac{x}{\alpha}\right) = 1 + A\frac{x^2}{\alpha^2}$ that

$$\begin{aligned} -\alpha\left\{g\left(g\left(-\tfrac{x}{\alpha}\right)\right)\right\} &= -\alpha\left\{1 + A\left(1 + A\tfrac{x^2}{\alpha^2}\right)^2\right\} \\ &= -\alpha\left\{1 + A + 2A^2\tfrac{x^2}{\alpha^2} + A^3\tfrac{x^4}{\alpha^4}\right\}. \end{aligned}$$

Thus one obtains

$$1 + Ax^2 = -\alpha(1 + A) - 2A^2\alpha\frac{x^2}{\alpha^2} - A^3\alpha\frac{x^4}{\alpha^4}.$$

Therefore one finds $1 = -\alpha(1 + A)$ and $A = -2A^2\frac{1}{\alpha}$. Then it follows that $2 = -\alpha(2 - \alpha)$ and hence

$$\alpha = 1 + \sqrt{3} = 2.73\ldots.$$

Problem 6. Suppose $n = 1$ first. Then there are two stable solutions $x_{2\pm}$ and two unstable solutions x_0 and x_- for $R_1 < R < R_2$.

Let $R_{n-1} < R < R_n$ next. Suppose there are 2^n fixed points of $L_R^{2n-1}(x) = x$, 2^{n-1} of which are stable periodic points of $L_R(x)$, while the remaining 2^{n-1} points are unstable periodic points. Then the former is destablized at $R = R_n$ and, instead, $2^{n-1} \times 2 = 2^n$ stable periodic orbits are produced while $2^{n-1} + 2^{n-1} = 2^n$ unstable periodic points result.

Problem 7. Observe that

$$\frac{\mathrm{d}f^n(x)}{\mathrm{d}x} = \frac{\mathrm{d}f^n(x)}{\mathrm{d}f^{n-1}(x)} \frac{\mathrm{d}f^{n-1}(x)}{\mathrm{d}x}$$

$$\frac{\mathrm{d}^2 f^n(x)}{\mathrm{d}x^2} = \frac{\mathrm{d}^2 f^n(x)}{\mathrm{d}(f^{n-1}(x))^2} \left(\frac{\mathrm{d}f^{n-1}(x)}{\mathrm{d}x} \right)^2 + \frac{\mathrm{d}f^n(x)}{\mathrm{d}f^{n-1}(x)} \frac{\mathrm{d}^2 f^{n-1}(x)}{\mathrm{d}x^2}$$

$$\frac{\mathrm{d}^3 f^n(x)}{\mathrm{d}x^3} = \frac{\mathrm{d}^3 f^n(x)}{\mathrm{d}(f^{n-1}(x))^3} \left(\frac{\mathrm{d}f^{n-1}(x)}{\mathrm{d}x} \right)^3 + 3 \frac{\mathrm{d}^2 f^n(x)}{\mathrm{d}(f^{n-1}(x))^2} \frac{\mathrm{d}f^{n-1}(x)}{\mathrm{d}x} \frac{\mathrm{d}^2 f^{n-1}(x)}{\mathrm{d}x^2} + \frac{\mathrm{d}f^n(x)}{\mathrm{d}f^{n-1}(x)} \frac{\mathrm{d}^3 f^{n-1}(x)}{\mathrm{d}x^3}.$$

It follows from the above that

$$\begin{aligned}
S[f^n(x)] &= \left(\frac{\mathrm{d}f^{n-1}(x)}{\mathrm{d}x} \right)^2 \frac{\dfrac{\mathrm{d}^3 f^n(x)}{\mathrm{d}(f^{n-1}(x))^3}}{\dfrac{\mathrm{d}f^n(x)}{\mathrm{d}f^{n-1}(x)}} + 3 \frac{\dfrac{\mathrm{d}^2 f^n(x)}{\mathrm{d}(f^{n-1}(x))^2} \dfrac{\mathrm{d}^2 f^{n-1}(x)}{\mathrm{d}x^2}}{\dfrac{\mathrm{d}f^n(x)}{\mathrm{d}f^{n-1}(x)}} \\
&\quad + \frac{\dfrac{\mathrm{d}^3 f^{n-1}(x)}{\mathrm{d}x^3}}{\dfrac{\mathrm{d}f^{n-1}(x)}{\mathrm{d}x}} - \frac{3}{2} \left(\frac{\dfrac{\mathrm{d}^2 f^n(x)}{\mathrm{d}(f^{n-1}(x))^2} \dfrac{\mathrm{d}f^{n-1}(x)}{\mathrm{d}x}}{\dfrac{\mathrm{d}f^n(x)}{\mathrm{d}f^{n-1}(x)}} + \frac{\dfrac{\mathrm{d}^2 f^{n-1}(x)}{\mathrm{d}x^2}}{\dfrac{\mathrm{d}f^{n-1}(x)}{\mathrm{d}x}} \right)^2 \\
&= \left(\frac{\mathrm{d}f^{n-1}(x)}{\mathrm{d}x} \right)^2 \left\{ \frac{\dfrac{\mathrm{d}^3 f(f^{n-1}(x))}{\mathrm{d}(f^{n-1}(x))^3}}{\dfrac{\mathrm{d}f(f^{n-1}(x))}{\mathrm{d}f^{n-1}(x)}} - \frac{3}{2} \left(\frac{\dfrac{\mathrm{d}^2 f(f^{n-1}(x))}{\mathrm{d}(f^{n-1}(x))^2}}{\dfrac{\mathrm{d}f(f^{n-1}(x))}{\mathrm{d}f^{n-1}(x)}} \right)^2 \right\} \\
&\quad + \frac{\dfrac{\mathrm{d}^3 f^{n-1}(x)}{\mathrm{d}x^3}}{\dfrac{\mathrm{d}f^{n-1}(x)}{\mathrm{d}x}} - \frac{3}{2} \left(\frac{\dfrac{\mathrm{d}^2 f^{n-1}(x)}{\mathrm{d}x^2}}{\dfrac{\mathrm{d}f^{n-1}(x)}{\mathrm{d}x}} \right)^2 \\
&= \left(\frac{\mathrm{d}f^{n-1}}{\mathrm{d}x} \right)^2 S[f]_{x=f^{n-1}(x)} + S[f^{n-1}]_{x=x}.
\end{aligned}$$

Problem 8.

(i) It follows from

$$f'(x) = nx^{n-1} \quad f''(x) = n(n-1)x^{n-2} \quad f'''(x) = n(n-1)(n-2)x^{n-3}$$

that

$$S[f] = \frac{n^2 - 1}{2x^2} \qquad (x \neq 0).$$

(ii) When $x < \frac{1}{2}$, one finds

$$S[f] = -\frac{(p^2 - 1)}{(2x - 1)^2}$$

from $f'(x) = 2pA(1 - 2x)^{p-1}$, $f''(x) = -4p(p - 1)A(1 - 2x)^{p-2}$ and $f'''(x) = 8p(p - 1)(p - 2)A(1 - 2x)^{p-3}$. One obtains the same result for $x > \frac{1}{2}$.

Problem 9. Observe that

$$
\begin{aligned}
-2\sqrt{f'}\frac{d^2}{dx^2}\left(\frac{1}{\sqrt{f'}}\right) &= -2\sqrt{f'}\frac{d}{dx}\left(-\frac{1}{2}\frac{f''}{\sqrt{(f')^3}}\right) = \sqrt{f'}\frac{d}{dx}\left(\frac{f''}{\sqrt{(f')^3}}\right) \\
&= \sqrt{f'}\frac{f'''(f')^{3/2} - \frac{3}{2}(f')^{1/2}(f'')^2}{(f')^3} \\
&= \frac{f'''}{f'} - \frac{3}{2}\left(\frac{f''}{f'}\right)^2 = S[f].
\end{aligned}
$$

Problem 10. Let α, β and γ be solutions of $f(x) = L_R^3(x) - x = 0$. It follows from

$$f'(x) = \frac{dL_R^3(x)}{dL_R^2(x)}\frac{dL_R^2(x)}{dL_R(x)}\frac{dL_R(x)}{dx} - 1$$

that

$$f'(\alpha) = f'(\beta) = f'(\gamma) = L_R{}'(\alpha)L_R{}'(\beta)L_R{}'(\gamma) - 1.$$

Thus $f'(\alpha) = 0$ implies $f'(\beta) = f'(\gamma) = 0$.

Problem 11. It follows from

$$
\begin{aligned}
L_R^3\left(\frac{1}{2}\right) &= RL_R^2\left(\frac{1}{2}\right)\left(1 - L_R^2\left(\frac{1}{2}\right)\right) \\
&= RRL_R\left(\frac{1}{2}\right)\left(1 - L_R\left(\frac{1}{2}\right)\right)\left\{1 - RL_R\left(\frac{1}{2}\right)\left(1 - L_R\left(\frac{1}{2}\right)\right)\right\}
\end{aligned}
$$

and $L_R\left(\frac{1}{2}\right) = R\frac{1}{2}\left(1 - \frac{1}{2}\right) = \frac{R}{4}$ that

$$L_R^3\left(\frac{1}{2}\right) = R^2\frac{R}{4}\left(1 - \frac{R}{4}\right)\left\{1 - R\frac{R}{4}\left(1 - \frac{R}{4}\right)\right\} = \frac{1}{2}.$$

This is simplified as

$$R^7 - 8R^6 + 16R^5 + 16R^4 - 64R^3 + 128 = 0.$$

Problem 12. Substitute $p = 3, 5, 11, 23$ into $\frac{2^{p-1} - 1}{p}$ to find $1, 3, 93, 182\,361$, respectively.

Problem 13. One finds from equation (3.21) that

$$\int_{x_1}^{x_2} \frac{\mathrm{d}x}{ax^2 + \varepsilon} = \int_{n_1}^{n_2} \mathrm{d}n.$$

Let $x = \sqrt{\frac{\varepsilon}{a}} \tan\theta$. Then

$$\frac{\mathrm{d}x}{ax^2 + \varepsilon} = \frac{1}{\sqrt{\varepsilon a}} \mathrm{d}\theta$$

and accordingly

$$\frac{1}{\sqrt{\varepsilon a}} \int_{\theta_1}^{\theta_2} \mathrm{d}\theta = \int_{n_1}^{n_2} \mathrm{d}n$$

from which one proves equation (3.22), where $\theta_1 = \tan^{-1} \sqrt{\frac{a}{\varepsilon}} x_1$ and $\theta_2 = \tan^{-1} \sqrt{\frac{a}{\varepsilon}} x_2$.

Problem 14. An equation introduced in appendix 3A is used. By putting

$$\begin{aligned} L_R^3(x) - x &= x(Rx + 1 - R)R^6(x - \alpha)^2(x - \beta)^2(x - \gamma)^2 \\ &= c_0(R) + c_1(x - \beta) + c_2(x - \beta)^2 + \ldots \end{aligned}$$

it is easily seen that equation (3.29) becomes

$$c_0 = 0 \quad c_1 = 0 \quad c_2(R) = \beta(R\beta + 1 - R)R^6(\beta - \alpha)^2(\beta - \gamma)^2.$$

On the other hand, equation (3.31) is the variation of the extremum of $L_R^3(x) - x$ with respect to R and hence

$$\begin{aligned} \frac{\mathrm{d}}{\mathrm{d}R} c_0(R) &= \frac{\mathrm{d}}{\mathrm{d}R}(L_R^3(\beta) - \beta) = \frac{\mathrm{d}}{\mathrm{d}R} L_R^3(\beta) \\ &= \frac{\mathrm{d}}{\mathrm{d}R} L_R^3(\beta) - \frac{7}{R}(L_R^3(\beta) - \beta) \qquad (*) \\ &= \frac{1}{R} \left\{(2R + 3)R^4\beta^2(\beta - 1)^2 + 2R^6\beta^3(\beta - 1)^3 + 4R^3\beta(\beta - 1) + 7\beta\right\} \\ &= \frac{1}{R} \left\{(2R + 3)R^2\gamma^2 - 2R^3\gamma^3 - 4R^2\gamma + 7\beta\right\} \end{aligned}$$

$$= \frac{1}{R}\left\{2R^3\gamma^2(1-\gamma)+3R^2\gamma(\gamma-1)-R^2\gamma+7\beta\right\}$$
$$= \frac{1}{R}\left\{2R^2\alpha-3R\alpha-R^2\gamma+7\beta\right\} = -\frac{\varepsilon}{d}.$$

The minus sign in front of $\frac{\varepsilon}{d}$ appears since d is defined by the equation $R = R_c - d$. We remark that the computation of the part (*) has been carried out with symbolic computation software on a personal computer.

Chapter 4

Problem 1. Substitute

$$x = A\cos\omega_0 t + B\sin\omega_0 t$$
$$y = m\dot{x} = m\omega_0(-A\sin\omega_0 t + B\cos\omega_0 t)$$

into $E = \frac{1}{2m}y^2 + \frac{1}{2}kx^2$ and use $\omega_0 = \sqrt{\frac{k}{m}}$ to find

$$E = \frac{k}{2}(A^2+B^2).$$

Problem 2. An addition of trigonometric functions yields $x = \sqrt{A^2+B^2}\sin(\omega_0 t+\delta)$. On the other hand, one obtains

$$y = m\dot{x} = m\omega_0\sqrt{A^2+B^2}\cos(\omega_0 t+\delta).$$

This represents an ellipse

$$\frac{x^2}{A^2+B^2} + \frac{y^2}{m^2\omega_0^2(A^2+B^2)} = 1$$

with an area S given by

$$S = \pi\sqrt{A^2+B^2}\,m\omega_0\sqrt{A^2+B^2} = \pi m\omega_0(A^2+B^2).$$

Therefore

$$\frac{E}{S} = \frac{k}{2\pi m\omega_0} = \frac{\omega_0}{2\pi} = V_0.$$

Problem 3. Let x_1 and x_2 be the fundamental solutions of equation (4.12). Then

$$\ddot{x}_1 + 2r\dot{x}_1 + \omega_0^2 x_1 = 0 \qquad (1)$$
$$\ddot{x}_2 + 2r\dot{x}_2 + \omega_0^2 x_2 = 0 \qquad (2)$$

By calculating (1) $\times x_2$ − (2) $\times x_1$ one obtains

$$\ddot{x}_1 x_2 - \ddot{x}_2 x_1 + 2r(\dot{x}_1 x_2 - \dot{x}_2 x_1) = 0.$$

It follows from (4.11) and (4.11′) that

$$\frac{\mathrm{d}}{\mathrm{d}t} W(x_1, x_2) + 2r W(x_1, x_2) = 0.$$

This is readily solved to yield

$$W(x_1, x_2) = W_0\, \mathrm{e}^{-2rt}.$$

Problem 4. Let $x = A\cos\omega t + B\sin\omega t$ and $y = \dot{x} = -\omega A\sin\omega t + \omega B\cos\omega t$. Then

$$A = \frac{\begin{vmatrix} x & \sin\omega t \\ y & \omega\cos\omega t \end{vmatrix}}{\begin{vmatrix} \cos\omega t & \sin\omega t \\ -\omega\sin\omega t & \omega\cos\omega t \end{vmatrix}} = \frac{1}{\omega}(x\omega\cos\omega t - y\sin\omega t)$$

$$B = \frac{1}{\omega}\begin{vmatrix} \cos\omega t & x \\ -\omega\sin\omega t & y \end{vmatrix} = \frac{1}{\omega}(y\cos\omega t + x\omega\sin\omega t).$$

Thus substituting

$$\frac{\partial \dot{x}}{\partial A} = -\omega\sin\omega t \qquad \frac{\partial A}{\partial x} = \cos\omega t$$

$$\frac{\partial \dot{x}}{\partial B} = \omega\cos\omega t \qquad \frac{\partial B}{\partial x} = \sin\omega t$$

into $\frac{\partial \dot{x}}{\partial x} = \frac{\partial \dot{x}}{\partial A}\frac{\partial A}{\partial x} + \frac{\partial \dot{x}}{\partial B}\frac{\partial B}{\partial x}$, one obtains

$$\frac{\partial \dot{x}}{\partial x} = -\omega\sin\omega t\cos\omega t + \omega\cos\omega t\sin\omega t = 0.$$

From

$$\dot{y} = -\omega^2 A\cos\omega t - \omega^2 B\sin\omega t$$

one obtains

$$\frac{\partial \dot{y}}{\partial A} = -\omega^2\cos\omega t \qquad \frac{\partial A}{\partial y} = -\frac{1}{\omega}\sin\omega t$$

$$\frac{\partial \dot{y}}{\partial B} = -\omega^2\sin\omega t \qquad \frac{\partial B}{\partial y} = \frac{1}{\omega}\cos\omega t.$$

Substituting them into $\frac{\partial \dot{y}}{\partial y} = \frac{\partial \dot{y}}{\partial A}\frac{\partial A}{\partial y} + \frac{\partial \dot{y}}{\partial B}\frac{\partial B}{\partial y}$, one finds

$$\frac{\partial \dot{y}}{\partial y} = (-\omega^2\cos\omega t)\left(-\frac{1}{\omega}\sin\omega t\right) + (-\omega^2\sin\omega t)\left(-\frac{1}{\omega}\cos\omega t\right) = 0.$$

Therefore it follows from $\frac{\dot{S}}{S} = \frac{\partial \dot{x}}{\partial x} + \frac{\partial \dot{y}}{\partial y}$ that $\dot{S}(t) = 0$.

Problem 5. Using $ma_r = F_r = mg\cos\theta - kx$, $ma_\theta = F_\theta = -mg\sin\theta$, $r = l + x$, $\dot{r} = \dot{x}$ and $\ddot{r} = \ddot{x}$, one obtains

$$m(\ddot{x} - (l+x)\dot{\theta}^2) = mg\cos\theta - kx \quad m((l+x)\ddot{\theta} + 2\dot{x}\dot{\theta}) = -mg\sin\theta$$

namely, equations (4.23).

Problem 6. Explicit evaluation of equations (4.26) yields

$$\begin{aligned}
\dot{x} &= \frac{\partial H}{\partial y} = \frac{y}{m} \\
\dot{y} &= -\frac{\partial H}{\partial x} = \frac{1}{m}\frac{\alpha^2}{(l+x)^3} + mg\cos\theta - kx \\
&= m(l+x)\dot{\theta}^2 + mg\cos\theta - kx \\
\dot{\theta} &= \frac{\partial H}{\partial \alpha} = \frac{1}{m}\frac{\alpha}{(l+x)^2} \\
\dot{\alpha} &= -\frac{\partial H}{\partial \theta} = -mg(l+x)\sin\theta.
\end{aligned}$$

It follows from these equations that

$$2m(l+x)\dot{x}\dot{\theta} + m(l+x)^2\ddot{\theta} = -mg(l+x)\sin\theta$$

$$(l+x)\ddot{\theta} + 2\dot{x}\dot{\theta} = -g\sin\theta.$$

Thus equations (4.26) yield two identities and the equation of motion (4.23). It also follows from the above equations that

$$\frac{\partial \dot{x}}{\partial x} = \frac{\partial \dot{y}}{\partial y} = \frac{\partial \dot{\theta}}{\partial \theta} = \frac{\partial \dot{\alpha}}{\partial \alpha} = 0.$$

Therefore one concludes that

$$\frac{\dot{V}}{V} = 0.$$

Problem 7. Substitution of $x = e^{\lambda t}$ into equation (4.12) yields

$$\lambda^2 + 2r\lambda + \omega_0 = 0. \tag{1}$$

Let λ_1 and λ_2 be the solutions of (1). It may be assumed that $\lambda_1 \neq \lambda_2$ so that the general solution is given by

$$x(t) = A\,e^{\lambda_1 t} + B\,e^{\lambda_2 t} \tag{2}$$

from which one obtains

$$\dot{x}(t) = A\lambda_1 e^{\lambda_1 t} + B\lambda_2 e^{\lambda_2 t}. \tag{3}$$

Note also that

$$\begin{aligned} x(t+\tau) &= A\,e^{\lambda_1(t+\tau)} + B\,e^{\lambda_2(t+\tau)} \\ &= A\,e^{\lambda_1 t}\,e^{\lambda_1 \tau} + B\,e^{\lambda_2 t}\,e^{\lambda_2 \tau}. \end{aligned} \tag{4}$$

$A\,e^{\lambda_1 t}$ and $B\,e^{\lambda_2 t}$ are found from (2) and (3) as

$$A\,e^{\lambda_1 t} = \frac{\lambda_2 x(t) - \dot{x}(t)}{\lambda_2 - \lambda_1} \qquad B\,e^{\lambda_2 t} = \frac{\dot{x}(t) - \lambda_1 x(t)}{\lambda_2 - \lambda_1}.$$

Substitution of these results into equation (4) yields

$$x(t+\tau) = \frac{\lambda_1 e^{\lambda_2 \tau} - \lambda_2 e^{\lambda_1 \tau}}{\lambda_1 - \lambda_2} x(t) + \frac{e^{\lambda_1 \tau} - e^{\lambda_2 \tau}}{\lambda_1 - \lambda_2} \dot{x}(t).$$

The solution is found if $\lambda_1 = -r + i\tilde{\omega}$ and $\lambda_2 = -r - i\tilde{\omega}$ are substituted.

An alternative solution is obtained if $x(t) = C\,e^{-rt}\sin(\tilde{\omega}t + \delta)$ is substituted into equation (4.12) directly.

Problem 8. Let us consider a rectangle with edges l and m. Suppose this is covered with squares with edge ε. The most efficient way of covering is achieved when squares mutually touch as shown in figure 4.20. The diameter of a square is $\sqrt{2}\,\varepsilon$ and the total number of them is

$$\left(\left[\frac{l}{\varepsilon}\right] + 1\right)\left(\left[\frac{m}{\varepsilon}\right] + 1\right) \simeq \frac{lm}{\varepsilon^2}.$$

Therefore

$$\inf \sum_k \varepsilon_k^\alpha = \frac{lm}{\varepsilon^2}(\sqrt{2}\varepsilon)^\alpha = 2^{\alpha/2} lm \varepsilon^{\alpha-2}.$$

This takes a finite fixed value $2lm$ as $\varepsilon \to 0$ when $\alpha = 2$. Therefore the Hausdorff dimension D_0 is equal to 2. It can be shown similarly that the Hausdorff dimension of a rectangular parellelpiped is 3.

Problem 9. Observe that $\varepsilon = l, n(\varepsilon) = 1$ for L_0; $\varepsilon = l/2, n(\varepsilon) = 4$ for L_1; $\varepsilon = l/4, n(\varepsilon) = 16$ for L_2 and hence $\varepsilon = l/2^n, n(\varepsilon) = 2^{2n}$ for L_n. Thus

$$D_{\mathrm{ca}} = \lim_{\varepsilon \to 0} \frac{\log n(\varepsilon)}{\log \frac{1}{\varepsilon}} = \lim_{n \to \infty} \frac{\log 2^{2n}}{\log \frac{l}{2^n}} = 2.$$

Problem 10.

$$D_q = \lim_{\varepsilon \to 0} \frac{1}{\log \varepsilon} \left. \frac{\frac{d}{dq}\left(\log \sum_{i=1}^{n(\varepsilon)} p_i^q\right)}{\frac{d}{dq}(q-1)} \right|_{q=1} = \lim_{\varepsilon \to 0} \frac{1}{\log \varepsilon} \left. \frac{\sum_{i=1}^{n(\varepsilon)} (\log p_i) p_i^q}{\sum_{i=1}^{n(\varepsilon)} p_i^q} \right|_{q=1}$$

$$= \lim_{\varepsilon \to 0} \frac{\sum_{i=1}^{n(\varepsilon)} (\log p_i) p_i}{\log \varepsilon}.$$

Problem 11. Let us consider the step C_2 for example. Then $n(\varepsilon) = 3$ and $\varepsilon = \frac{1}{3}$ in equation (4.33). Let $p_1 = \frac{1}{2}$, $p_2 = 0$ and $p_3 = \frac{1}{2}$ to find

$$\frac{1}{q-1} \frac{\log\left(2\left(\frac{1}{2}\right)^q\right)}{\log \frac{1}{3}} = \frac{\log 2}{\log 3}.$$

Similarly $n(\varepsilon) = 9 = 3^2$ and $\varepsilon = \frac{1}{3^2}$ for C_3. It follows from $p_1 = p_3 = p_7 = p_9 = \frac{1}{2^2}$ that

$$\frac{1}{q-1} \frac{\log\left(2^2\left(\frac{1}{2^2}\right)^q\right)}{\log \frac{1}{3^2}} = \frac{\log 2}{\log 3}.$$

The right-hand side remains the same up to C_∞ and hence $D_q = \frac{\log 2}{\log 3}$.
Problem 12. If one puts $f(x) = \log x$ $(x > 0)$ and $a_i = x_i = p_i$ $(i = 1, 2, \ldots, n(\varepsilon))$ one obtains

$$\sum_{i=1}^{n(\varepsilon)} p_i \log p_i \le \log\left(\sum_{i=1}^{n(\varepsilon)} p_i^2\right).$$

Thus it follows that $D_2 \le D_1$.

Let us put $f(x) = -x \log x$ $(0 < x \le 1)$, $a_i = \frac{1}{n(\varepsilon)}$, $x_i = p_i (i = 1, 2, \ldots, n(\varepsilon))$ next to find

$$\sum p_i \log p_i \ge -\log n(\varepsilon).$$

Then one finds $D_1 \le D_0$.
Problem 13. Since $p^{(n)} = p^n = (l_1^m l_2^{n-m})^\alpha$ $(m = 0, 1, 2, \ldots, n)$, one obtains

$$\alpha = \frac{n \log p}{m \log l_1 + (n-m) \log l_2} = \frac{\log p}{X \log l_1 + (1-X) \log l_2}.$$

On the other hand, one has

$$N = (l_1^m l_2^{n-m})^{-f(\alpha)} \simeq \frac{n^n}{m^m (n-m)^{n-m}}$$

from which one obtains

$$-f(\alpha) = \frac{n \log n - m \log m - (n-m) \log(n-m)}{m \log l_1 + (n-m) \log l_2}.$$

Therefore

$$f(\alpha) = \frac{X \log X + (1 - X)\log(1 - X)}{X \log l_1 + (1 - X)\log l_2}.$$

From $p = \frac{1}{2}, l_1 = \frac{1}{4}, l_2 = \frac{1}{2}$, one obtains

$$\alpha = \frac{-\log 2}{-2X\log 2 - (1 - X)\log 2} = \frac{1}{1 + X}$$

$$f(\alpha) = \frac{X \log X + (1 - X)\log(1 - X)}{-2X\log 2 - (1 - X)\log 2}$$

$$= -\frac{1}{\log 2}\frac{X \log X + (1 - X)\log(1 - X)}{1 + X}$$

$$= -\frac{1}{\log 2}\{(1 - \alpha)\log(1 - \alpha) + (2\alpha - 1)\log(2\alpha - 1) - \alpha \log \alpha\}.$$

The figure below is a schematic graph showing the relation between α and $f(\alpha)$.

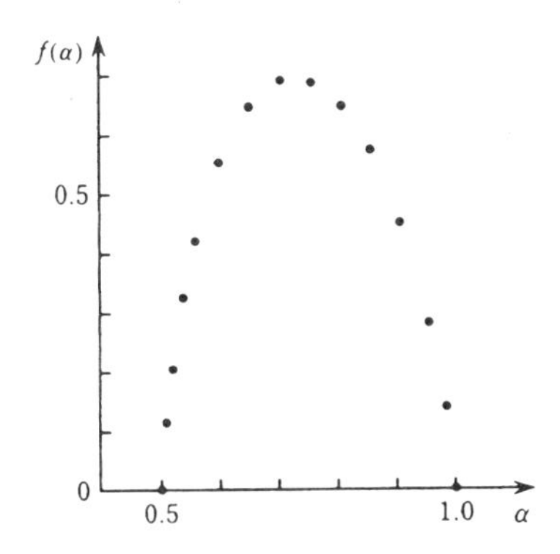

Problem 14. It follows from equation (4.71) that

$$(q - 1)D_q = q\alpha(q) - f(\alpha(q))$$

$$\frac{\mathrm{d}}{\mathrm{d}q}[(q - 1)D_q] = \alpha(q) + q\frac{\mathrm{d}\alpha(q)}{\mathrm{d}q} - \frac{\mathrm{d}\alpha(q)}{\mathrm{d}q}f'(\alpha(q)).$$

Equation (4.73) follows since $f'(\alpha(q)) = q$ (see equation (4.70)).
Problem 15. The gamma function is defined by

$$\Gamma(n + 1) = \int_0^\infty \mathrm{e}^{-x}x^n \,\mathrm{d}x = \int_0^\infty \mathrm{e}^{-x+n\log x}\,\mathrm{d}x.$$

Let us put $g(x) = x - n\log x$, from which one derives

$$g'(x) = 1 - \frac{n}{x} \qquad g''(x) = \frac{n}{x^2}.$$

One obtains $x_0 = n$ from $g'(x_0) = 0$. Expansion of $g(x)$ around x_0 yields

$$g(x) = n - n \log n + \frac{1}{2!}\frac{1}{n}(x - n)^2 + \dots .$$

Use the equation

$$I \simeq \sqrt{\frac{2\pi}{g''(x_0)}} f(x_0)\, \mathrm{e}^{-g(x_0)}$$

in appendix 4B with $f(x_0) \equiv 1$ to obtain

$$I = \sqrt{2\pi n}\, \mathrm{e}^{-(n - n \log n)} \simeq \sqrt{2\pi n} \left(\frac{n}{\mathrm{e}}\right)^n .$$

We note that this approximation yields

$$n = 1;\ \ I = 0.922\dots\ \ (1! = 1)$$
$$n = 10;\ I = 3598\,695.6\ (10! = 3628\,800)$$

with relative error of about $\frac{1}{12n}$.

Problem 16. We have

$$\alpha = \frac{\log p}{X \log l_1 + (1 - X) \log l_2}$$
$$f(\alpha) = \frac{X \log X + (1 - X) \log(1 - X)}{X \log l_1 + (1 - X) \log l_2}.$$

Let us put

$$p = \frac{1}{2} \qquad l_1 = \frac{1}{\alpha_{\mathrm{PD}}} \qquad l_2 = \frac{1}{\alpha_{\mathrm{PD}}^2} \qquad \alpha_{\mathrm{PD}} = 2.502\,807\,876\dots$$

to draw the figure below. Compare this with Fig. 4.35.

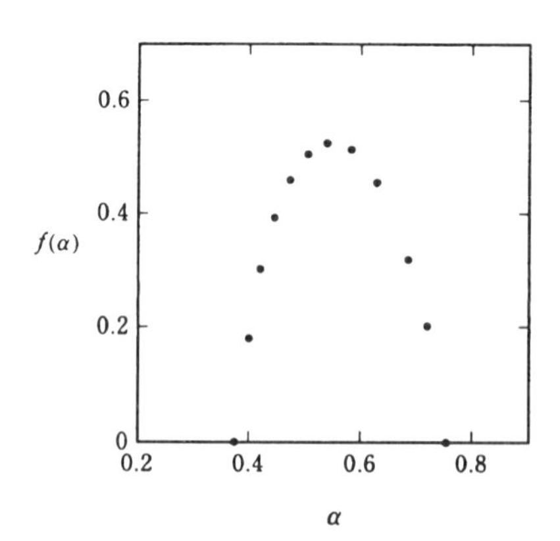

Index